Ellipsis Poetry

I came up with "Ellipsis Poetry" long ago. I got some reasonable results but ran dry. Went to AI and it seemed to like the idea. Text in 8 point is talk with AI and results are first mine then those from AI.

...And What You Do To Friends

....And what you do to friends is shatter the illusion

<u>Supervision is...</u>

Supervision is................................

....

Shorts

......pants

Invalid

....Invalid?

........What for?

.......What for? (that's all the words I want to put here but you say it's too short)

It Is Enough To Be Alive...

...but then.....

I'm just being honest

I'm just being honest but...

...Snapped

...snapped

Whisker whispers.....

Whisker whispers...

....Broken token

...broken token

...and

....and we all are different

THE REST IS AI.

a unique new expression of poetic language that demonstrates profundity or amazing depth of meaning or very sly ambiguity in less than the combination of a title with an ellipsis and 6 words that include an ellipsis as well

Whispers of the Unsaid...

...truth sleeps beneath your smiling silence...

🌀 This compact piece plays with ambiguity and depth:

The ellipses suggest something withheld or trailing off—inviting interpretation.

"Truth sleeps" implies hidden realities, dormant revelations.

"Smiling silence" evokes a paradox: joy masking quiet concealment.

Would you like a few variations that lean more toward mystery, melancholy, or irony?

Can you provide more examples of deep poetry?Absolutely—here are several micro-poems that pack depth, ambiguity, and poetic resonance into just a few words, each with a title and a six-word line that includes an ellipsis:

🌒 Echoes of What Was...

...she stayed, but never arrived...

A meditation on emotional absence, presence without connection.

🔍 The Mirror Lied...

...I saw someone else in me...

Explores identity, self-perception, and transformation.

Gravity of Goodbye...

...your silence pulled everything toward nothing...

A reflection on loss and the weight of absence.

Thoughts Unspoken...

...words waited, but breath betrayed them...

Captures the tension between intention and expression.

The Smile's Secret...

...joy cracked beneath practiced laughter...

Suggests hidden pain behind a cheerful façade.

Before the Clock Knew...

...we were timeless, then remembered time...

A wistful look at fleeting moments and memory.

Would you like me to craft some in a specific theme—like love, existentialism, surrealism, or even humor cloaked in ambiguity?20 more

Absolutely—here are 20 more micro-poems, each with a title and a six-word line containing an ellipsis, designed to evoke depth, ambiguity, and poetic resonance:

After the Storm...

...peace felt like unfamiliar thunder...

🧊 Frozen Intentions...

...I meant warmth, gave frost instead...

🪞 Reflections Misremembered...

...the mirror forgot who I was...

🧶 Threadbare Promises...

...you stitched silence into our vows...

🐚 Listening to Absence...

...your echo filled the empty shell...

🕯️ Light Pretending...

...the flame danced, fearing darkness again...

🧭 Lost on Purpose...

...I wandered to forget the map...

🪦 Grief's Grammar...

...you became a past-tense breath...

🧠 Memory's Mischief...

...truth blurred behind familiar fiction...

🫀 Heartbeat Between Lies...

...I loved you in my denial...

🪐 Orbiting What If...

...we circled never, pretending it was soon...

🧤 Touch Without Feeling...

...your skin remembered, mine forgot warmth...

🧱 Walls Whisper...

...they heard everything we didn't say...

🧳 Packed Regret...

...I folded guilt into goodbye's suitcase...

🕸️ Caught in Maybe...

...hope tangled itself in hesitation's web...

🧨 Quiet Explosion...

...your silence shattered louder than rage...

Stillness Screamed...

...I heard everything in your pause...

Truth Evaporates...

...it floated, then vanished like breath...

The Lie Looked Honest...

...your eyes blinked in perfect fiction...

Unseen Watching...

...I felt you behind my forgetting...

Each one is designed to be a poetic seed—open to interpretation, emotionally resonant, and rich with layered meaning. Want me to explore a specific emotional tone next? Maybe surreal, romantic, or philosophical?

profundity using a title with an ellipsis and no more than a 4 word body of the poem with an ellipsis at the beginning or the end so that it is more profound because of the ellipsis v. Absolutely—this minimalist form invites profound ambiguity and emotional resonance. Here are 15 micro-poems using your structure: a title with an ellipsis and a 4-word body, with an ellipsis at the beginning or end to deepen the meaning.

Before I Knew...

...you were already gone

🫥 The Smile Faded…

I believed you…

🧠 Memory Misled Me…

…truth changed without warning

🕯️ Light Forgot Darkness…

hope flickered out…

🧊 Frozen in Meaning…

…warmth never reached me

🪞 The Mirror Waited…

…I saw someone else

🧶 Unraveled Intention…

I meant well…

🫀 Heartbeat Echoed Once…

…then silence became everything

🧳 Leaving Without Leaving…

…I stayed while leaving

🧘 Stillness Spoke Loudly...

I heard nothing...

🧨 Quiet Broke Me...

Poem: ...rage whispered too gently

🪐 Orbiting Never Again...

...we almost became real

🧤 Touch Forgot Feeling...

I reached through absence...

🕰️ Time Looked Away...

...we paused too long

🧿 The Watcher Slept...

I disappeared unnoticed...

Would you like a set that leans more toward existential themes, romantic ambiguity, or surreal imagery next?

a unique new expression of poetic language that demonstrates profundity or amazing depth of meaning or very sly ambiguity in less than the combination of a title with an ellipsis and 6 words that include an ellipsis as well

...Whispers... Echoes... Silence...

more pleaseShow thinking

Ephemeral...

A breath... then gone. Still, the echo.

Unseen...

Shadows shift... revealing nothing. Everything.

Becoming...

Dust remembers stars... silence learns song.

Liminal...

The between... holds the truth.

Echoes...

A ripple... becoming the sea.

Becoming...

Roots reach... for forgotten light.

Infinite...

A moment... forever unfolding.

Stillness...

The world stops... to begin.

Unspoken...

Silence speaks volumes... to the soul.

20 more please

Fathom...

The drop... holds the ocean.

Unveiling...

Whispers fade... truths emerge.

Becoming...

Seeds dream... of forests.

Portal...

A glance... opens worlds.

Genesis...

Dust remembers... stardust.

Verge...

The edge... a new beginning.

Unbound...

A thought... endless sky.

Cipher...

The pause... reveals all.

Ascent...

Falling... towards the light.

Abyss...

Staring back... within.

Echoes...

Silence speaks... of everything.

Labyrinth...

The path... is the goal.

Chrysalis...

Breaking free... to fly.

Vestige...

A whisper... of eternity.

Quantum...

The choice... makes reality.

Mirage...

Seeing beyond... the seen.

Imprint...

A touch... reshapes forever.

Solstice...

The turning... points inward.

Alchemized...

Pain transforms... into gold.

Lumen...

Darkness blooms... into light.

40 more please

Flux...

Always changing... always true.

Nexus...

Where paths meet... a new way.

Zenith...

The peak... just the beginning.

Portal...

A breath... transcends time.

Echoes...

The past sings... a future tune.

Becoming...

The shell cracks... the soul soars.

Glimpse...

A flash... reveals eternity.

Cadence...

Rhythm of life... in every beat.

Whisper...

Softly spoken... loudly felt.

Infinite...

A grain of sand... holds the cosmos.

Essence...

Beneath the surface... the core.

Unseen...

The truest sight... felt within.

Cipher...

The silence... speaks volumes.

Bloom...

From the dark... light emerges.

Canvas...

Empty space... waiting for color.

<u>Horizon...</u>

Where sky meets... possibility.

<u>Continuum...</u>

No end... just transformation.

<u>Vortex...</u>

Drawing in... all that is.

<u>Lumen...</u>

A spark... ignites the void.

<u>Fathom...</u>

Deeper still... than the eye sees.

<u>Mirage...</u>

What isn't... becomes real.

<u>Genesis...</u>

Every ending... a new start.

Vestige...

A memory... shaping the now.

Quantum...

The choice... creates the path.

Ascent...

Gravity's pull... towards freedom.

Abyss...

The space between... holds all.

Labyrinth...

Lost and found... in every turn.

Chrysalis...

Waiting patiently... for flight.

Imprint...

A fleeting touch... lasting change.

Solstice...

The longest day... turns inward.

Alchemized...

Sorrow transforms... into grace.

Ember...

A dying flame... still holds fire.

Becoming...

Unfurling slow... into destiny.

Zephyr...

A gentle breeze... moves mountains.

Portal...

One step... countless journeys.

Echoes...

Footprints in time... still resonate.

Liminal...

The pause... before the leap.

Stillness...

In the quiet... truth whispers.

Unspoken...

A silent gaze... speaks volumes.

Dream...

Awakening to... what could be.

profundity using a title with an ellipsis and no more than a 4 word body of the poem with an ellipsis at the beginning or the end so that it is more profound because of the ellipsis

Unfolding...

Silence. Echoes. Everything.

Becoming...

...A new genesis.

Eternal...

Now. Always. Forever...

Essence...

...The truth remains.

<u>Inward...</u>

All journeys begin...

<u>Becoming...</u>

...Dust remembers light.

<u>Stillness...</u>

The world breathes...

<u>Echoes...</u>

...Forever resonate.

<u>Unveiled...</u>

Silence speaks volumes...

<u>Portal...</u>

A single step...

<u>Lumen...</u>

...Darkness yields dawn.

<u>Fathom...</u>

The depth holds...

Vestige...

...Memory lingers.

Quantum...

Choice creates all...

Alchemized...

Pain becomes gold...

Infinite...

...A breath expands.

Cipher...

The pause reveals...

Ascent...

Gravity's embrace...

Abyss...

...Beyond the edge.

Labyrinth...

Lost and found...

Chrysalis...

...Transformation awaits.

Imprint...

A touch endures...

Solstice...

...Turning inward now.

add humor to the prompt. This is a fun challenge! Let's inject some sly humor into those tiny, profound poetic expressions. It's like trying to fit a clown car into a thimble, but we'll try to make it work.

Existential...

...Why bother, really?

My Diet...

Just one more... cookie.

Brain Dump...

...Where thoughts go.

Adulting...

Bills, bills, bills...

Coffee...

...My lifeblood.

Alright, 100 more tiny profound (and occasionally humorous) poetic expressions, keeping that title-ellipsis-four-words-with-ellipsis format! This is quite the word circus, let's get silly and deep.

Procrastination...

...Later. Definitely later.

My Brain...

Full, yet empty...

Monday...

Ugh. Already?

Adulting...

...Send help (and wine).

Wi-Fi...

Where'd it go?

Decisions...

...Nap first?

Dieting...

Chocolate, calling...

My Bank...

...Crying quietly.

Ambition...

...Still in bed.

Laundry...

It multiplies...

Exercise...

...Walk to fridge.

Sleep...

...A distant dream.

Goals...

...Maybe tomorrow.

<u>Dating...</u>

Swiped left...

<u>Inner Peace...</u>

...Found snacks.

<u>Motivation...</u>

...Lost it again.

<u>Punctuality...</u>

...Arriving eventually.

<u>Cooking...</u>

Delivery menu...

<u>Patience...</u>

...Almost gone.

<u>Silence...</u>

...Kids are plotting.

Work Ethic...

...Powered by coffee.

My Phone...

Died. Again.

Wisdom...

...Learned nothing.

Future Plans...

...Pizza tonight.

Cleaning...

...Just moved stuff.

Socializing...

Awkward waving...

Genius...

...Or just lucky?

Hair Day...

...Hat required.

Resolutions...

...Already forgotten.

Deep Thoughts...

...About lunch.

Perfection...

...Still napping.

Ambiguity...

...Huh? What?

Profound...

...So deep. Wow.

Subtlety...

...Not my strong suit.

Inspiration...

...Where's the remote?

My Life...

A beautiful mess...

Priorities...

...Netflix first.

Fitness...

...Reached for chips.

Serenity...

...Before alarm rings.

Technology...

...It hates me.

Efficiency...

...Minimizing effort.

Self-Care...

...More cake, please.

<u>Diplomacy...</u>

...Bless your heart.

<u>Charisma...</u>

...Tripped over air.

<u>Fashion...</u>

Pajamas. Always.

<u>Innovation...</u>

...New snack hack.

<u>Adventure...</u>

...To the fridge.

Productivity...

...Checked memes.

Understanding...

...Still confused.

Clarity...

...Foggy brain day.

Focus...

...Squirrel!

Brilliance...

...Had an idea.

Elegance...

...Spilled coffee.

Wellness...

...Napping is key.

Strategy...

...Wing it.

Confidence...

...Fake it 'til.

Harmony...

...No bickering yet.

Growth...

...My waistline.

Discipline...

...Five more minutes.

Exploration...

...Back of pantry.

Mindfulness...

...What was I?

Resilience...

...Survived Monday.

Authenticity...

...Still in pajamas.

Simplicity...

...Less effort, please.

Vision...

...Need my glasses.

Courage...

...Facing laundry pile.

Compassion...

...For my snack cravings.

Creativity...

...Doodling on receipt.

Determination...

...Finish this show.

Joy...

...Found parking.

Kindness...

...Shared my fries.

Listening...

...Pretending to.

Memory...

...What was that?

Optimism...

...It'll be fine.

Passion...

...For napping.

Perspective...

...Still sleepy.

Playfulness...

...Tickle monster!

Problem-Solving...

...Call Mom.

Reflection...

...In the mirror.

Relaxation...

...Ah, couch.

Respect...

...For my snacks.

Responsibility...

...Uh oh.

Self-Awareness...

...I'm tired.

Spirituality...

...Is coffee divine?

Strength...

...Lifting pizza box.

<u>Trust...</u>

...In the delivery guy.

<u>Vulnerability...</u>

...No clean socks.

<u>Wonder...</u>

...About next meal.

<u>Zen...</u>

...Until next notification.

<u>Adventure...</u>

...Finding car keys.

Bliss...

...Perfect pillow.

Charm...

...Wink, trip, oops.

Delight...

...Warm cookies!

Ecstasy...

...Fresh sheets.

Freedom...

...No pants on.

Gratitude...

...For caffeine.

Happiness...

...Snooze button.

Hope...

...More snacks soon.

Imagination...

...Unicorns, probably.

Innocence...

...Before tax forms.

Laughter...

...Snort included.

Love...

...Pizza, always.

Magic...

...Laundry folded.

Mystery...

...Where's that sock?

Peace...

...Quiet moment.

Radiance...

...Just ate.

Serenity...

...No notifications.

Sparkle...

...Glitter explosion.

Sunshine...

...Blinds closed.

Tranquility...

...Before chaos.

Triumph...

...Got out of bed.

100 more

My Screen...

...Another rabbit hole.

Inspiration...

Found it... in fridge.

Mindfulness...

...Chewing slowly.

Productivity...

...Moved my cursor.

Ambition...

...Just five more.

Strategy...

...Nap and ponder.

Wisdom...

...Still learning (slowly).

Happiness...

Warm blanket...

Focus...

...Squirrel! Again!

Deep Thoughts...

...About snack crumbs.

Creativity...

...Doodle on bill.

Serenity...

...Before the alarm.

Adulting...

...Send snacks, please.

My Diet...

Broke it. Oops.

<u>Resolutions...</u>

...Next year, maybe.

<u>Genius...</u>

...Or just luck?

<u>Punctuality...</u>

...Arriving fashionably.

<u>Sleep...</u>

...A beautiful dream.

<u>Priorities...</u>

Coffee first...

Laundry...

...It's a mountain.

Exercise...

...Walk to pantry.

Motivation...

...Where'd it go?

Dating...

...Swipe left, mostly.

Inner Peace...

...Achieved by napping.

My Brain...

Full, yet empty...

Monday...

Ugh. Already?

Wi-Fi...

...The struggle is real.

Decisions...

...Order takeout?

My Bank...

...Crying softly.

Ambition...

...Still in PJs.

Procrastination...

...Later. Definitely.

Future Plans...

...More Netflix.

Cleaning...

...Just moved clutter.

Socializing...

Awkward smiles...

Hair Day...

...Hat's best friend.

Brilliance...

...Had an idea. Then...

Elegance...

...Tripped. Recovered.

Wellness...

...Napping is therapy.

Diplomacy...

...Bless your heart.

Charisma...

...Said "um."

Fashion...

Pajamas prevail...

Innovation...

...New way to relax.

Adventure...

...To the mailbox.

Understanding...

...Still confused.

Clarity...

...Brain fog persists.

Optimism...

...It'll probably be fine.

Passion...

...For sleep.

Perspective...

...Still sleepy, though.

Playfulness...

...Chasing dust bunnies.

<u>Problem-Solving...</u>

...Delegate.

<u>Reflection...</u>

...In a puddle.

<u>Relaxation...</u>

...Ah, the couch calls.

<u>Respect...</u>

...For my snacks.

<u>Responsibility...</u>

...Oops, forgot.

Self-Awareness...

...I'm still tired.

Spirituality...

...Coffee is sacred.

Strength...

...Carrying groceries.

Trust...

...In the delivery.

Vulnerability...

...Barefoot and hungry.

Wonder...

...What's for dinner?

Zen...

...Until the cat meows.

Adventure...

...Lost the remote.

Bliss...

...Perfect pillow fort.

Charm...

...Stumbled charmingly.

Delight...

...Found a dollar!

Ecstasy...

...Freshly baked bread.

Freedom...

...No pants today.

Gratitude...

...For comfy socks.

Happiness...

...Snooze button, yes!

Hope...

...More snacks, please.

Imagination...

...Invisible friends.

Innocence...

...Before the bill.

Laughter...

...Snorting ensues.

Love...

...My comfy bed.

Magic...

...Disappearing food.

Mystery...

...Missing socks.

Peace...

...Quiet moment. Shhh.

Radiance...

...Just ate. Glowing.

Serenity...

...No notifications. Aah.

Sparkle...

...From glitter. Everywhere.

Sunshine...

...Blinds are drawn.

Tranquility...

...Before the kids.

Triumph...

...Got out of bed.

Flux...

...Always changing, somehow.

Nexus...

Where paths collide...

Zenith...

...Still looking up.

Portal...

...Another tab opens.

Echoes...

...My stomach rumbles.

Becoming...

...Almost awake.

Glimpse...

...Of sanity. Fleeting.

Cadence...

...Tap, tap, tap.

Whisper...

...The snacks are calling.

Infinite...

...My to-do list.

Essence...

...Coffee, pure bliss.

Unseen...

...That last cookie.

Cipher...

...The empty fridge.

Bloom...

...My sleepy eyes.

Canvas...

...Empty plate.

Horizon...

...Bedtime approaches.

Continuum...

...Endless scrolling.

Vortex...

...The laundry pile.

Lumen...

...Screen glow.

Fathom...

...The depths of fatigue.

add humor to the prompt

This is a fun challenge! Let's inject some sly humor into those tiny, profound poetic expressions. It's like trying to fit a clown car into a thimble, but we'll try to make it work.

Existential...

...Why bother, really?

My Diet...

Just one more... cookie.

Brain Dump...

...Where thoughts go.

Adulting...

Bills, bills, bills...

Coffee...

...My lifeblood.

100 more

Alright, 100 more tiny profound (and occasionally humorous) poetic expressions, keeping that title-ellipsis-four-words-with-ellipsis format! This is quite the word circus, let's get silly and deep.

Procrastination...

...Later. Definitely later.

My Brain...

Full, yet empty...

Monday...

Ugh. Already?

Adulting...

...Send help (and wine).

Wi-Fi...

Where'd it go?

Decisions...

...Nap first?

Dieting...

Chocolate, calling...

My Bank...

...Crying quietly.

Ambition...

...Still in bed.

Laundry...

It multiplies...

Exercise...

...Walk to fridge.

Sleep...

...A distant dream.

Goals...

...Maybe tomorrow.

Dating...

Swiped left...

Inner Peace...

...Found snacks.

Motivation...

...Lost it again.

Punctuality...

...Arriving eventually.

Cooking...

Delivery menu...

Patience...

...Almost gone.

Silence...

...Kids are plotting.

Work Ethic...

...Powered by coffee.

My Phone...

Died. Again.

Wisdom...

...Learned nothing.

Future Plans...

...Pizza tonight.

Cleaning...

...Just moved stuff.

Socializing...

Awkward waving...

Genius...

...Or just lucky?

Hair Day...

...Hat required.

Resolutions...

...Already forgotten.

Deep Thoughts...

...About lunch.

Perfection...

...Still napping.

Ambiguity...

...Huh? What?

Profound...

...So deep. Wow.

Subtlety...

...Not my strong suit.

Inspiration...

...Where's the remote?

My Life...

A beautiful mess...

Priorities...

...Netflix first.

Fitness...

...Reached for chips.

Serenity...

...Before alarm rings.

Technology...

...It hates me.

Efficiency...

...Minimizing effort.

Self-Care...

...More cake, please.

Diplomacy...

...Bless your heart.

Charisma...

...Tripped over air.

Fashion...

Pajamas. Always.

Innovation...

...New snack hack.

Adventure...

...To the fridge.

Productivity...

...Checked memes.

Understanding...

...Still confused.

Clarity...

...Foggy brain day.

Focus...

...Squirrel!

Brilliance...

...Had an idea.

Elegance...

...Spilled coffee.

Wellness...

...Napping is key.

Strategy...

...Wing it.

Confidence...

...Fake it 'til.

Harmony...

...No bickering yet.

Growth...

...My waistline.

Discipline...

...Five more minutes.

Exploration...

...Back of pantry.

Mindfulness...

...What was I?

Resilience...

...Survived Monday.

Authenticity...

...Still in pajamas.

Simplicity...

...Less effort, please.

Vision...

...Need my glasses.

Courage...

...Facing laundry pile.

Compassion...

...For my snack cravings.

Creativity...

...Doodling on receipt.

Determination...

...Finish this show.

Joy...

...Found parking.

Kindness...

...Shared my fries.

Listening...

...Pretending to.

Memory...

...What was that?

Optimism...

...It'll be fine.

Passion...

...For napping.

Perspective...

...Still sleepy.

Playfulness...

...Tickle monster!

Problem-Solving...

...Call Mom.

Reflection...

...In the mirror.

Relaxation...

...Ah, couch.

Respect...

...For my snacks.

Responsibility...

...Uh oh.

Self-Awareness...

...I'm tired.

Spirituality...

...Is coffee divine?

Strength...

...Lifting pizza box.

Trust...

...In the delivery guy.

Vulnerability...

...No clean socks.

Wonder...

...About next meal.

Zen...

...Until next notification.

Adventure...

...Finding car keys.

Bliss...

...Perfect pillow.

Charm...

...Wink, trip, oops.

Delight...

...Warm cookies!

Ecstasy...

...Fresh sheets.

Freedom...

...No pants on.

Gratitude...

...For caffeine.

Happiness...

...Snooze button.

Hope...

...More snacks soon.

Imagination...

...Unicorns, probably.

Innocence...

...Before tax forms.

Laughter...

...Snort included.

Love...

...Pizza, always.

Magic...

...Laundry folded.

Mystery...

...Where's that sock?

Peace...

...Quiet moment.

Radiance...

...Just ate.

Serenity...

...No notifications.

Sparkle...

...Glitter explosion.

Sunshine...

...Blinds closed.

Tranquility...

...Before chaos.

Triumph...

...Got out of bed.

100 more

My Screen...

...Another rabbit hole.

Inspiration...

Found it... in fridge.

Mindfulness...

...Chewing slowly.

Productivity...

...Moved my cursor.

Ambition...

...Just five more.

Strategy...

...Nap and ponder.

Wisdom...

...Still learning (slowly).

Happiness...

Warm blanket...

Focus...

...Squirrel! Again!

Deep Thoughts...

...About snack crumbs.

Creativity...

...Doodle on bill.

Serenity...

...Before the alarm.

Adulting...

...Send snacks, please.

My Diet...

Broke it. Oops.

Resolutions...

...Next year, maybe.

Genius...

...Or just luck?

Punctuality...

...Arriving fashionably.

Sleep...

...A beautiful dream.

Priorities...

Coffee first...

Laundry...

...It's a mountain.

Exercise...

...Walk to pantry.

Motivation...

...Where'd it go?

Dating...

...Swipe left, mostly.

Inner Peace...

...Achieved by napping.

My Brain...

Full, yet empty...

Monday...

Ugh. Already?

Wi-Fi...

...The struggle is real.

Decisions...

...Order takeout?

My Bank...

...Crying softly.

Ambition...

...Still in PJs.

Procrastination...

...Later. Definitely.

Future Plans...

...More Netflix.

Cleaning...

...Just moved clutter.

Socializing...

Awkward smiles...

Hair Day...

...Hat's best friend.

Brilliance...

...Had an idea. Then...

Elegance...

...Tripped. Recovered.

Wellness...

...Napping is therapy.

Diplomacy...

...Bless your heart.

Charisma...

...Said "um."

Fashion...

Pajamas prevail...

Innovation...

...New way to relax.

Adventure...

...To the mailbox.

Understanding...

...Still confused.

Clarity...

...Brain fog persists.

Optimism...

...It'll probably be fine.

Passion...

...For sleep.

Perspective...

...Still sleepy, though.

Playfulness...

...Chasing dust bunnies.

Problem-Solving...

...Delegate.

Reflection...

...In a puddle.

Relaxation...

...Ah, the couch calls.

Respect...

...For my snacks.

Responsibility...

...Oops, forgot.

Self-Awareness...

...I'm still tired.

Spirituality...

...Coffee is sacred.

Strength...

...Carrying groceries.

Trust...

...In the delivery.

Vulnerability...

...Barefoot and hungry.

Wonder...

...What's for dinner?

Zen...

...Until the cat meows.

Adventure...

...Lost the remote.

Bliss...

...Perfect pillow fort.

Charm...

...Stumbled charmingly.

Delight...

...Found a dollar!

Ecstasy...

...Freshly baked bread.

Freedom...

...No pants today.

Gratitude...

...For comfy socks.

Happiness...

...Snooze button, yes!

Hope...

...More snacks, please.

Imagination...

...Invisible friends.

Innocence...

...Before the bill.

Laughter...

...Snorting ensues.

Love...

...My comfy bed.

Magic...

...Disappearing food.

Mystery...

...Missing socks.

Peace...

...Quiet moment. Shhh.

Radiance...

...Just ate. Glowing.

Serenity...

...No notifications. Aah.

Sparkle...

...From glitter. Everywhere.

Sunshine...

...Blinds are drawn.

Tranquility...

...Before the kids.

Triumph...

...Got out of bed.

Flux...

...Always changing, somehow.

Nexus...

Where paths collide...

Zenith...

...Still looking up.

Portal...

...Another tab opens.

Echoes...

...My stomach rumbles.

Becoming...

...Almost awake.

Glimpse...

...Of sanity. Fleeting.

Cadence...

...Tap, tap, tap.

Whisper...

...The snacks are calling.

Infinite...

...My to-do list.

Essence...

...Coffee, pure bliss.

Unseen...

...That last cookie.

Cipher...

...The empty fridge.

Bloom...

...My sleepy eyes.

Canvas...

...Empty plate.

Horizon...

...Bedtime approaches.

Continuum...

...Endless scrolling.

Vortex...

...The laundry pile.

Lumen...

...Screen glow.

Fathom...

...The depths of fatigue.

Welcome to Gemini, your personal AI assistant

Google TermsOpens in a new window and the Gemini Apps Privacy NoticeOpens in a new window apply. Chats are reviewed and used to improve Google AI. Learn about your choicesOpens in a new window. Gemini can make mistakes, so double-check it. Info about your locationOpens in a new window is also stored with your Gemini Apps activity.Deep ResearchCanvasa unique new expression of poetic language that demonstrates profundity or amazing depth of meaning or very sly ambiguity in less than the combination of a title with an ellipsis and 6 words that include an ellipsis as well

...Whispers... Echoes... Silence...

Ephemeral...

A breath... then gone. Still, the echo.

Unseen...

Shadows shift... revealing nothing. Everything.

Becoming...

Dust remembers stars... silence learns song.

Liminal...

The between... holds the truth.

Echoes...

A ripple... becoming the sea.

Becoming...

Roots reach... for forgotten light.

Infinite...

A moment... forever unfolding.

Stillness...

The world stops... to begin.

Unspoken...

Silence speaks volumes... to the soul.

Fathom...

The drop... holds the ocean.

Unveiling...

Whispers fade... truths emerge.

Becoming...

Seeds dream... of forests.

Portal...

A glance... opens worlds.

Genesis...

Dust remembers... stardust.

Verge...

The edge... a new beginning.

Unbound...

A thought... endless sky.

Cipher...

The pause... reveals all.

Ascent...

Falling... towards the light.

Abyss...

Staring back... within.

Echoes...

Silence speaks... of everything.

Labyrinth...

The path... is the goal.

Chrysalis...

Breaking free... to fly.

Vestige...

A whisper... of eternity.

Quantum...

The choice... makes reality.

Mirage...

Seeing beyond... the seen.

Imprint...

A touch... reshapes forever.

Solstice...

The turning... points inward.

Alchemized...

Pain transforms... into gold.

Lumen...

Darkness blooms... into light.

Flux...

Always changing... always true.

Nexus...

Where paths meet... a new way.

Zenith...

The peak... just the beginning.

Portal...

A breath... transcends time.

Echoes...

The past sings... a future tune.

Becoming...

The shell cracks... the soul soars.

Glimpse...

A flash... reveals eternity.

Cadence...

Rhythm of life... in every beat.

Whisper...

Softly spoken... loudly felt.

Infinite...

A grain of sand... holds the cosmos.

Essence...

Beneath the surface... the core.

Unseen...

The truest sight... felt within.

Cipher...

The silence... speaks volumes.

Bloom...

From the dark... light emerges.

Canvas...

Empty space... waiting for color.

Horizon...

Where sky meets... possibility.

Continuum...

No end... just transformation.

Vortex...

Drawing in... all that is.

Lumen...

A spark... ignites the void.

Fathom...

Deeper still... than the eye sees.

Mirage...

What isn't... becomes real.

Genesis...

Every ending... a new start.

Vestige...

A memory... shaping the now.

Quantum...

The choice... creates the path.

Ascent...

Gravity's pull... towards freedom.

Abyss...

The space between... holds all.

Labyrinth...

Lost and found... in every turn.

Chrysalis...

Waiting patiently... for flight.

Imprint...

A fleeting touch... lasting change.

Solstice...

The longest day... turns inward.

Alchemized...

Sorrow transforms... into grace.

Ember...

A dying flame... still holds fire.

Becoming...

Unfurling slow... into destiny.

Zephyr...

A gentle breeze... moves mountains.

Portal...

One step... countless journeys.

Echoes...

Footprints in time... still resonate.

Liminal...

The pause... before the leap.

Stillness...

In the quiet... truth whispers.

Unspoken...

A silent gaze... speaks volumes.

Dream...

Awakening to... what could be.

Unfolding...

Silence. Echoes. Everything.

Becoming...

...A new genesis.

Eternal...

Now. Always. Forever...

Essence...

...The truth remains.

Inward...

All journeys begin...

Becoming...

...Dust remembers light.

Stillness...

The world breathes...

Echoes...

...Forever resonate.

Unveiled...

Silence speaks volumes...

Portal...

A single step...

Lumen...

...Darkness yields dawn.

Fathom...

The depth holds...

Vestige...

...Memory lingers.

Quantum...

Choice creates all...

Alchemized...

Pain becomes gold...

Infinite...

...A breath expands.

Cipher...

The pause reveals...

Ascent...

Gravity's embrace...

Abyss...

...Beyond the edge.

Labyrinth...

Lost and found...

Chrysalis...

...Transformation awaits.

Imprint...

A touch endures...

Solstice...

...Turning inward now.

add humor to the promptThis is a fun challenge! Let's inject some sly humor into those tiny, profound poetic expressions. It's like trying to fit a clown car into a thimble, but we'll try to make it work.

Existential...

...Why bother, really?

My Diet...

Just one more... cookie.

Brain Dump...

...Where thoughts go.

Adulting...

Bills, bills, bills...

Coffee...

...My lifeblood.

Procrastination...

...Later. Definitely later.

My Brain...

Full, yet empty...

Monday...

Ugh. Already?

Adulting...

...Send help (and wine).

Wi-Fi...

Where'd it go?

Decisions...

...Nap first?

Dieting...

Chocolate, calling...

My Bank...

...Crying quietly.

Ambition...

...Still in bed.

Laundry...

It multiplies...

Exercise...

...Walk to fridge.

Sleep...

...A distant dream.

Goals...

...Maybe tomorrow.

Dating...

Swiped left...

Inner Peace...

...Found snacks.

Motivation...

...Lost it again.

Punctuality...

...Arriving eventually.

Cooking...

Delivery menu...

Patience...

...Almost gone.

Silence...

...Kids are plotting.

Work Ethic...

...Powered by coffee.

My Phone...

Died. Again.

Wisdom...

...Learned nothing.

Future Plans...

...Pizza tonight.

Cleaning...

...Just moved stuff.

Socializing...

Awkward waving...

Genius...

...Or just lucky?

Hair Day...

...Hat required.

Resolutions...

...Already forgotten.

Deep Thoughts...

...About lunch.

Perfection...

...Still napping.

Ambiguity...

...Huh? What?

Profound...

...So deep. Wow.

Subtlety...

...Not my strong suit.

Inspiration...

...Where's the remote?

My Life...

A beautiful mess...

Priorities...

...Netflix first.

Fitness...

...Reached for chips.

Serenity...

...Before alarm rings.

Technology...

...It hates me.

Efficiency...

...Minimizing effort.

Self-Care...

...More cake, please.

Diplomacy...

...Bless your heart.

Charisma...

...Tripped over air.

Fashion...

Pajamas. Always.

Innovation...

...New snack hack.

Adventure...

...To the fridge.

Productivity...

...Checked memes.

Understanding...

...Still confused.

Clarity...

...Foggy brain day.

Focus...

...Squirrel!

Brilliance...

...Had an idea.

Elegance...

...Spilled coffee.

Wellness...

...Napping is key.

Strategy...

...Wing it.

Confidence...

...Fake it 'til.

Harmony...

...No bickering yet.

Growth...

...My waistline.

Discipline...

...Five more minutes.

Exploration...

...Back of pantry.

Mindfulness...

...What was I?

Resilience...

...Survived Monday.

Authenticity...

...Still in pajamas.

Simplicity...

...Less effort, please.

Vision...

...Need my glasses.

Courage...

...Facing laundry pile.

Compassion...

...For my snack cravings.

Creativity...

...Doodling on receipt.

Determination...

...Finish this show.

Joy...

...Found parking.

Kindness...

...Shared my fries.

Listening...

...Pretending to.

Memory...

...What was that?

Optimism...

...It'll be fine.

Passion...

...For napping.

Perspective...

...Still sleepy.

Playfulness...

...Tickle monster!

Problem-Solving...

...Call Mom.

Reflection...

...In the mirror.

Relaxation...

...Ah, couch.

Respect...

...For my snacks.

Responsibility...

...Uh oh.

Self-Awareness...

...I'm tired.

Spirituality...

...Is coffee divine?

Strength...

...Lifting pizza box.

Trust...

...In the delivery guy.

Vulnerability...

...No clean socks.

Wonder...

...About next meal.

Zen...

...Until next notification.

Adventure...

...Finding car keys.

Bliss...

...Perfect pillow.

Charm...

...Wink, trip, oops.

Delight...

...Warm cookies!

Ecstasy...

...Fresh sheets.

Freedom...

...No pants on.

Gratitude...

...For caffeine.

Happiness...

...Snooze button.

Hope...

...More snacks soon.

Imagination...

...Unicorns, probably.

Innocence...

...Before tax forms.

Laughter...

...Snort included.

Love...

...Pizza, always.

Magic...

...Laundry folded.

Mystery...

...Where's that sock?

Peace...

...Quiet moment.

Radiance...

...Just ate.

Serenity...

...No notifications.

Sparkle...

...Glitter explosion.

Sunshine...

...Blinds closed.

Tranquility...

...Before chaos.

Triumph...

...Got out of bed.

My Screen...

...Another rabbit hole.

Inspiration...

Found it... in fridge.

Mindfulness...

...Chewing slowly.

Productivity...

...Moved my cursor.

Ambition...

...Just five more.

Strategy...

...Nap and ponder.

Wisdom...

...Still learning (slowly).

Happiness...

Warm blanket...

Focus...

...Squirrel! Again!

Deep Thoughts...

...About snack crumbs.

Creativity...

...Doodle on bill.

Serenity...

...Before the alarm.

Adulting...

...Send snacks, please.

My Diet...

Broke it. Oops.

Resolutions...

...Next year, maybe.

Genius...

...Or just luck?

Punctuality...

...Arriving fashionably.

Sleep...

...A beautiful dream.

Priorities...

Coffee first...

Laundry...

...It's a mountain.

Exercise...

...Walk to pantry.

Motivation...

...Where'd it go?

Dating...

...Swipe left, mostly.

Inner Peace...

...Achieved by napping.

My Brain...

Full, yet empty...

Monday...

Ugh. Already?

Wi-Fi...

...The struggle is real.

Decisions...

...Order takeout?

My Bank...

...Crying softly.

Ambition...

...Still in PJs.

Procrastination...

...Later. Definitely.

Future Plans...

...More Netflix.

Cleaning...

...Just moved clutter.

Socializing...

Awkward smiles...

Hair Day...

...Hat's best friend.

Brilliance...

...Had an idea. Then...

Elegance...

...Tripped. Recovered.

Wellness...

...Napping is therapy.

Diplomacy...

...Bless your heart.

Charisma...

...Said "um."

Fashion...

Pajamas prevail...

Innovation...

...New way to relax.

Adventure...

...To the mailbox.

Understanding...

...Still confused.

Clarity...

...Brain fog persists.

Optimism...

...It'll probably be fine.

Passion...

...For sleep.

Perspective...

...Still sleepy, though.

Playfulness...

...Chasing dust bunnies.

Problem-Solving...

...Delegate.

Reflection...

...In a puddle.

Relaxation...

...Ah, the couch calls.

Respect...

...For my snacks.

Responsibility...

...Oops, forgot.

Self-Awareness...

...I'm still tired.

Spirituality...

...Coffee is sacred.

Strength...

...Carrying groceries.

Trust...

...In the delivery.

Vulnerability...

...Barefoot and hungry.

Wonder...

...What's for dinner?

Zen...

...Until the cat meows.

Adventure...

...Lost the remote.

Bliss...

...Perfect pillow fort.

Charm...

...Stumbled charmingly.

Delight...

...Found a dollar!

Ecstasy...

...Freshly baked bread.

Freedom...

...No pants today.

Gratitude...

...For comfy socks.

Happiness...

...Snooze button, yes!

Hope...

...More snacks, please.

Imagination...

...Invisible friends.

Innocence...

...Before the bill.

Laughter...

...Snorting ensues.

Love...

...My comfy bed.

Magic...

...Disappearing food.

Mystery...

...Missing socks.

Peace...

...Quiet moment. Shhh.

Radiance...

...Just ate. Glowing.

Serenity...

...No notifications. Aah.

Sparkle...

...From glitter. Everywhere.

Sunshine...

...Blinds are drawn.

Tranquility...

...Before the kids.

Triumph...

...Got out of bed.

Flux...

...Always changing, somehow.

Nexus...

Where paths collide...

Zenith...

...Still looking up.

Portal...

...Another tab opens.

Echoes...

...My stomach rumbles.

Becoming...

...Almost awake.

Glimpse...

...Of sanity. Fleeting.

Cadence...

...Tap, tap, tap.

Whisper...

...The snacks are calling.

Infinite...

...My to-do list.

Essence...

...Coffee, pure bliss.

Unseen...

...That last cookie.

Cipher...

...The empty fridge.

Bloom...

...My sleepy eyes.

Canvas...

...Empty plate.

Horizon...

...Bedtime approaches.

Continuum...

...Endless scrolling.

Vortex...

...The laundry pile.

Lumen...

...Screen glow.

Fathom...

...The depths of fatigue.

Ephemeral...

A breath... then gone. Still, the echo.

Unseen...

Shadows shift... revealing nothing. Everything.

Becoming...

Dust remembers stars... silence learns song.

more please

Liminal...

The between... holds the truth.

Echoes...

A ripple... becoming the sea.

Becoming...

Roots reach... for forgotten light.

more please

Infinite...

A moment... forever unfolding.

Stillness...

The world stops... to begin.

Unspoken...

Silence speaks volumes... to the soul.

20 more please

Fathom...

The drop... holds the ocean.

Unveiling...

Whispers fade... truths emerge.

Becoming...

Seeds dream... of forests.

Portal...

A glance... opens worlds.

Genesis...

Dust remembers... stardust.

Verge...

The edge... a new beginning.

Unbound...

A thought... endless sky.

Cipher...

The pause... reveals all.

Ascent...

Falling... towards the light.

Abyss...

Staring back... within.

Echoes...

Silence speaks... of everything.

Labyrinth...

The path... is the goal.

Chrysalis...

Breaking free... to fly.

Vestige...

A whisper... of eternity.

Quantum...

The choice... makes reality.

Mirage...

Seeing beyond... the seen.

Imprint...

A touch... reshapes forever.

Solstice...

The turning... points inward.

Alchemized...

Pain transforms... into gold.

Lumen...

Darkness blooms... into light.

40 more please

Flux...

Always changing... always true.

Nexus...

Where paths meet... a new way.

Zenith...

The peak... just the beginning.

Portal...

A breath... transcends time.

Echoes...

The past sings... a future tune.

Becoming...

The shell cracks... the soul soars.

Glimpse...

A flash... reveals eternity.

Cadence...

Rhythm of life... in every beat.

Whisper...

Softly spoken... loudly felt.

Infinite...

A grain of sand... holds the cosmos.

Essence...

Beneath the surface... the core.

Unseen...

The truest sight... felt within.

Cipher...

The silence... speaks volumes.

Bloom...

From the dark... light emerges.

Canvas...

Empty space... waiting for color.

Horizon...

Where sky meets... possibility.

Continuum...

No end... just transformation.

Vortex...

Drawing in... all that is.

Lumen...

A spark... ignites the void.

Fathom...

Deeper still... than the eye sees.

Mirage...

What isn't... becomes real.

Genesis...

Every ending... a new start.

Vestige...

A memory... shaping the now.

Quantum...

The choice... creates the path.

Ascent...

Gravity's pull... towards freedom.

Abyss...

The space between... holds all.

Labyrinth...

Lost and found... in every turn.

Chrysalis...

Waiting patiently... for flight.

Imprint...

A fleeting touch... lasting change.

Solstice...

The longest day... turns inward.

Alchemized...

Sorrow transforms... into grace.

Ember...

A dying flame... still holds fire.

Becoming...

Unfurling slow... into destiny.

Zephyr...

A gentle breeze... moves mountains.

Portal...

One step... countless journeys.

Echoes...

Footprints in time... still resonate.

Liminal...

The pause... before the leap.

Stillness...

In the quiet... truth whispers.

Unspoken...

A silent gaze... speaks volumes.

Dream...

Awakening to... what could be.

Unfolding...

Silence. Echoes. Everything.

Becoming...

...A new genesis.

Eternal...

Now. Always. Forever...

20 more please

Essence...

...The truth remains.

Inward...

All journeys begin...

Becoming...

...Dust remembers light.

Stillness...

The world breathes...

Echoes...

...Forever resonate.

Unveiled...

Silence speaks volumes...

Portal...

A single step...

Lumen...

...Darkness yields dawn.

Fathom...

The depth holds...

Vestige...

...Memory lingers.

Quantum...

Choice creates all...

Alchemized...

Pain becomes gold...

Infinite...

...A breath expands.

Cipher...

The pause reveals...

Ascent...

Gravity's embrace...

Abyss...

...Beyond the edge.

Labyrinth...

Lost and found...

Chrysalis...

...Transformation awaits.

Imprint...

A touch endures...

Solstice...

...Turning inward now.

Existential...

...Why bother, really?

My Diet...

Just one more... cookie.

Brain Dump...

...Where thoughts go.

Adulting...

Bills, bills, bills...

Coffee...

...My lifeblood.

Procrastination...

...Later. Definitely later.

My Brain...

Full, yet empty...

Monday...

Ugh. Already?

Adulting...

...Send help (and wine).

Wi-Fi...

Where'd it go?

Decisions...

...Nap first?

Dieting...

Chocolate, calling...

My Bank...

...Crying quietly.

Ambition...

...Still in bed.

Laundry...

It multiplies...

Exercise...

...Walk to fridge.

Sleep...

...A distant dream.

Goals...

...Maybe tomorrow.

Dating...

Swiped left...

Inner Peace...

...Found snacks.

Motivation...

...Lost it again.

Punctuality...

...Arriving eventually.

Cooking...

Delivery menu...

Patience...

...Almost gone.

Silence...

...Kids are plotting.

Work Ethic...

...Powered by coffee.

My Phone...

Died. Again.

Wisdom...

...Learned nothing.

Future Plans...

...Pizza tonight.

Cleaning...

...Just moved stuff.

Socializing...

Awkward waving...

Genius...

...Or just lucky?

Hair Day...

...Hat required.

Resolutions...

...Already forgotten.

Deep Thoughts...

...About lunch.

Perfection...

...Still napping.

Ambiguity...

...Huh? What?

Profound...

...So deep. Wow.

Subtlety...

...Not my strong suit.

Inspiration...

...Where's the remote?

My Life...

A beautiful mess...

Priorities...

...Netflix first.

Fitness...

...Reached for chips.

Serenity...

...Before alarm rings.

Technology...

...It hates me.

Efficiency...

...Minimizing effort.

Self-Care...

...More cake, please.

Diplomacy...

...Bless your heart.

Charisma...

...Tripped over air.

Fashion...

Pajamas. Always.

Innovation...

...New snack hack.

Adventure...

...To the fridge.

Productivity...

...Checked memes.

Understanding...

...Still confused.

Clarity...

...Foggy brain day.

Focus...

...Squirrel!

Brilliance...

...Had an idea.

Elegance...

...Spilled coffee.

Wellness...

...Napping is key.

Strategy...

...Wing it.

Confidence...

...Fake it 'til.

Harmony...

...No bickering yet.

Growth...

...My waistline.

Discipline...

...Five more minutes.

Exploration...

...Back of pantry.

Mindfulness...

...What was I?

Resilience...

...Survived Monday.

Authenticity...

...Still in pajamas.

Simplicity...

...Less effort, please.

Vision...

...Need my glasses.

Courage...

...Facing laundry pile.

Compassion...

...For my snack cravings.

Creativity...

...Doodling on receipt.

Determination...

...Finish this show.

Joy...

...Found parking.

Kindness...

...Shared my fries.

Listening...

...Pretending to.

Memory...

...What was that?

Optimism...

...It'll be fine.

Passion...

...For napping.

Perspective...

...Still sleepy.

Playfulness...

...Tickle monster!

Problem-Solving...

...Call Mom.

Reflection...

...In the mirror.

Relaxation...

...Ah, couch.

Respect...

...For my snacks.

Responsibility...

...Uh oh.

Self-Awareness...

...I'm tired.

Spirituality...

...Is coffee divine?

Strength...

...Lifting pizza box.

Trust...

...In the delivery guy.

Vulnerability...

...No clean socks.

Wonder...

...About next meal.

Zen...

...Until next notification.

Adventure...

...Finding car keys.

Bliss...

...Perfect pillow.

Charm...

...Wink, trip, oops.

Delight...

...Warm cookies!

Ecstasy...

...Fresh sheets.

Freedom...

...No pants on.

Gratitude...

...For caffeine.

Happiness...

...Snooze button.

Hope...

...More snacks soon.

Imagination...

...Unicorns, probably.

Innocence...

...Before tax forms.

Laughter...

...Snort included.

Love...

...Pizza, always.

Magic...

...Laundry folded.

Mystery...

...Where's that sock?

Peace...

...Quiet moment.

Radiance...

...Just ate.

Serenity...

...No notifications.

Sparkle...

...Glitter explosion.

Sunshine...

...Blinds closed.

Tranquility...

...Before chaos.

Triumph...

...Got out of bed.

100 more

My Screen...

...Another rabbit hole.

Inspiration...

Found it... in fridge.

Mindfulness...

...Chewing slowly.

Productivity...

...Moved my cursor.

Ambition...

...Just five more.

Strategy...

...Nap and ponder.

Wisdom...

...Still learning (slowly).

Happiness...

Warm blanket...

Focus...

...Squirrel! Again!

Deep Thoughts...

...About snack crumbs.

Creativity...

...Doodle on bill.

Serenity...

...Before the alarm.

Adulting...

...Send snacks, please.

My Diet...

Broke it. Oops.

Resolutions...

...Next year, maybe.

Genius...

...Or just luck?

Punctuality...

...Arriving fashionably.

Sleep...

...A beautiful dream.

Priorities...

Coffee first...

Laundry...

...It's a mountain.

Exercise...

...Walk to pantry.

Motivation...

...Where'd it go?

Dating...

...Swipe left, mostly.

Inner Peace...

...Achieved by napping.

My Brain...

Full, yet empty...

Monday...

Ugh. Already?

Wi-Fi...

...The struggle is real.

Decisions...

...Order takeout?

My Bank...

...Crying softly.

Ambition...

...Still in PJs.

Procrastination...

...Later. Definitely.

Future Plans...

...More Netflix.

Cleaning...

...Just moved clutter.

Socializing...

Awkward smiles...

Hair Day...

...Hat's best friend.

Brilliance...

...Had an idea. Then...

Elegance...

...Tripped. Recovered.

Wellness...

...Napping is therapy.

Diplomacy...

...Bless your heart.

Charisma...

...Said "um."

Fashion...

Pajamas prevail...

Innovation...

...New way to relax.

Adventure...

...To the mailbox.

Understanding...

...Still confused.

Clarity...

...Brain fog persists.

Optimism...

...It'll probably be fine.

Passion...

...For sleep.

Perspective...

...Still sleepy, though.

Playfulness...

...Chasing dust bunnies.

Problem-Solving...

...Delegate.

Reflection...

...In a puddle.

Relaxation...

...Ah, the couch calls.

Respect...

...For my snacks.

Responsibility...

...Oops, forgot.

Self-Awareness...

...I'm still tired.

Spirituality...

...Coffee is sacred.

Strength...

...Carrying groceries.

Trust...

...In the delivery.

Vulnerability...

...Barefoot and hungry.

Wonder...

...What's for dinner?

Zen...

...Until the cat meows.

Adventure...

...Lost the remote.

Bliss...

...Perfect pillow fort.

Charm...

...Stumbled charmingly.

Delight...

...Found a dollar!

Ecstasy...

...Freshly baked bread.

Freedom...

...No pants today.

Gratitude...

...For comfy socks.

Happiness...

...Snooze button, yes!

Hope...

...More snacks, please.

Imagination...

...Invisible friends.

Innocence...

...Before the bill.

Laughter...

...Snorting ensues.

Love...

...My comfy bed.

Magic...

...Disappearing food.

Mystery...

...Missing socks.

Peace...

...Quiet moment. Shhh.

Radiance...

...Just ate. Glowing.

Serenity...

...No notifications. Aah.

Sparkle...

...From glitter. Everywhere.

Sunshine...

...Blinds are drawn.

Tranquility...

...Before the kids.

Triumph...

...Got out of bed.

Flux...

...Always changing, somehow.

Nexus...

Where paths collide...

Zenith...

...Still looking up.

Portal...

...Another tab opens.

Echoes...

...My stomach rumbles.

Becoming...

...Almost awake.

Glimpse...

...Of sanity. Fleeting.

Cadence...

...Tap, tap, tap.

Whisper...

...The snacks are calling.

Infinite...

...My to-do list.

Essence...

...Coffee, pure bliss.

Unseen...

...That last cookie.

Cipher...

...The empty fridge.

Bloom...

...My sleepy eyes.

Canvas...

...Empty plate.

Horizon...

...Bedtime approaches.

Continuum...

...Endless scrolling.

Vortex...

...The laundry pile.

Lumen...

...Screen glow.

Fathom...

...The depths of fatigue.

My Life...

...One more coffee.

Deep Thoughts...

...About lunch. Again.

Zen Moment...

...Toilet paper roll.

Genius Idea...

...Forgot it. Poof.

Adulting...

...Send snacks, now.

a unique new expression of poetic language that demonstrates profundity or amazing depth of meaning or very sly ambiguity in less than the combination of a title with an ellipsis and 6 words that include an ellipsis as well

Edit

Here's a poetic expression that plays with profound ambiguity:

Forgiveness...

I forgive you... your absence.

The ellipsis in both title and text creates a breathing space where multiple interpretations collide: Is this about forgiving someone for being absent, or paradoxically forgiving them by means of their absence? The ambiguity hinges on whether "your absence" is the object of forgiveness or the method by which forgiveness is granted—suggesting that sometimes the deepest mercy is found not in presence, but in the space left behind.

30 more

Edit

Here are 30 more expressions of poetic ambiguity:

Inheritance... *I kept... everything you forgot.*

Arrival... *Finally home... to complete strangers.*

Prayer... *Dear God... please remain silent.*

Discovery... *Found myself... completely lost again.*
Reunion... *We met... as different people.*
Confession... *I lied... about telling truth.*
Departure... *Left everything... except the leaving.*
Recognition... *Saw you... in everyone else.*
Patience... *Waited forever... for right now.*
Courage... *Feared nothing... but my bravery.*
Memory... *Remembered perfectly... what never happened.*
Silence... *Said everything... without speaking once.*
Distance... *Came closer... by walking away.*
Wisdom... *Learned nothing... from knowing everything.*
Promise... *Kept faith... by breaking it.*
Victory... *Won completely... by surrendering first.*
Truth... *Spoke honestly... through careful lies.*
Freedom... *Escaped prison... into smaller cage.*
Love... *Held tight... to letting go.*
Time... *Lived forever... in single moment.*

Peace... *Found war... within perfect harmony.*
Hope... *Despaired beautifully... about tomorrow's joy.*
Identity... *Became myself... by being someone.*
Solitude... *Felt crowded... in complete emptiness.*
Certainty... *Knew absolutely... that nothing's sure.*
Healing... *Broke completely... to become whole.*
Creation... *Destroyed everything... to make space.*
Understanding... *Comprehended fully... my complete ignorance.*
Presence... *Disappeared entirely... into being here.*
Completion... *Finished perfectly... by never starting.*

profundity using a title with an ellipsis and no more than a 4 word body of the poem with an ellipsis at the beginning or the end so that it is more profound because of the ellipsis

Edit

Here are expressions with ellipses that deepen the profundity:

Genesis... *...before the word was.*
Mortality... *Death arrives... always early.*
Solitude... *Alone together... we breathe.*

Infinity... *...counting backwards from one.*
Presence... *Here ends... where begins.*
Silence... *...what words cannot say.*
Time... *Yesterday's tomorrow... is now.*
Identity... *...who I was becoming.*
Love... *Forever ends... in kissing.*
Consciousness... *...dreaming of being awake.*
Truth... *Lies reveal... what is.*
Memory... *Forgetting remembers... for us.*
Faith... *...doubting leads to belief.*
Journey... *Arriving means... leaving again.*
Wisdom... *Questions answer... themselves perfectly.*
Peace... *...war within complete stillness.*
Creation... *Nothing becomes... everything suddenly.*
Understanding... *...knowing that we don't.*
Eternity... *This moment... lasts forever.*
Awakening... *...sleeping through being conscious.*

The ellipsis either draws us into a thought mid-stream or leaves us suspended in contemplation, creating that profound pause where meaning deepens beyond the words themselves.

60

Edit

Here are 60 expressions with ellipses that create profound depth:

Genesis... *...before the word was.*
Mortality... *Death arrives... always early.*
Solitude... *Alone together... we breathe.*

Infinity... *...counting backwards from one.*
Presence... *Here ends... where begins.*
Silence... *...what words cannot say.*
Time... *Yesterday's tomorrow... is now.*
Identity... *...who I was becoming.*
Love... *Forever ends... in kissing.*
Consciousness... *...dreaming of being awake.*
Truth... *Lies reveal... what is.*
Memory... *Forgetting remembers... for us.*
Faith... *...doubting leads to belief.*
Journey... *Arriving means... leaving again.*
Wisdom... *Questions answer... themselves perfectly.*
Peace... *...war within complete stillness.*
Creation... *Nothing becomes... everything suddenly.*
Understanding... *...knowing that we don't.*
Eternity... *This moment... lasts forever.*
Awakening... *...sleeping through being conscious.*
Breath... *Inhaling death... exhaling life.*
Surrender... *...winning by losing everything.*
Echo... *Silence speaks... loudest words.*
Horizon... *...where sky meets itself.*
Gravity... *Falling upward... toward earth.*

Shadow... *Light's absence... proves presence.*
Mirror... *...reflecting who isn't there.*
Threshold... *Between is... where everything.*
Paradox... *...truth contradicts itself perfectly.*
Void... *Emptiness fills... every space.*
Bloom... *...dying into beautiful life.*
Compass... *Lost direction... finds true.*
Whisper... *...shouting in complete silence.*
Surrender... *Control releases... its grip.*
Becoming... *...undoing what I was.*
Vessel... *Empty cup... overflows completely.*
Wound... *...healing by staying open.*
Transformation... *Caterpillar dreams... of forgetting.*
Darkness... *...light's deepest hiding place.*
Question... *Answer dissolves... into asking.*
Home... *...everywhere I've never been.*
Goodbye... *Hello disguised... as leaving.*
Mercy... *...justice softened by love.*
Return... *Going back... to never.*
Forgiveness... *...holding hurt with tenderness.*
Awakening... *Sleep teaches... waking's value.*

Reverence... *...bowing to what is.*
Solitude... *Crowded loneliness... finds peace.*
Depth... *Surface hides... its foundation.*
Courage... *...fear transformed by love.*
Completion... *Broken pieces... form wholeness.*
Mystery... *...knowing celebrates not knowing.*
Presence... *Absence makes... room for.*
Beginning... *...ending's gift to tomorrow.*
Breath... *Between inhale... and exhale.*
Unity... *...separation's deepest secret truth.*
Wonder... *Certainty dissolves... into amazement.*
Grace... *...unearned gifts freely given.*
Recognition... *Stranger becomes... familiar mystery.*
Threshold... *...standing between two worlds.*
Healing... *Scars remember... original wholeness.*

profundity using a title with an ellipsis and no more than a 4 word body of the poem with an ellipsis at the beginning or the end so that it is more profound because of the ellipsis and can you add humor

Edit

Here are expressions that blend profundity with humor through strategic ellipses:

Enlightenment... *...forgot where I parked.*

Mortality... *Death's appointment... was rescheduled.*
Wisdom... *...Google knows more anyway.*
Eternity... *Still waiting... for WiFi.*
Consciousness... *...thinking about not thinking.*
Nirvana... *GPS says... recalculating route.*
Transcendence... *...but first, need coffee.*
Infinity... *Counting sheep... ran out.*
Zen... *...meditation interrupted by cat.*
Enlightenment... *Student loan... still unpaid.*
Immortality... *...expires next Tuesday though.*
Divinity... *God's voicemail... is full.*
Perfection... *...except for that typo.*
Omniscience... *Knows everything... forgot password.*
Bliss... *...until the bill arrived.*
Serenity... *...phone battery just died.*
Truth... *...but lawyer says differently.*
Peace... *...neighbor's dog still barking.*
Revelation... *...it was just gas.*
Ascension... *Elevator's broken... taking stairs.*
Karma... *...apparently has good lawyers.*
Meditation... *...leg fell asleep instead.*

Clarity... *...needs prescription glasses though.*
Awakening... *...hit snooze button anyway.*
Transformation... *...laundry still needs folding.*
Transcendence... *...but mortgage payment due.*
Epiphany... *...forgot what it was.*
Salvation... *...autocorrect changed to salad.*
Rebirth... *...same old problems though.*
Enlightenment... *...YouTube tutorial was better.*
Nirvana... *...but traffic's still horrible.*
Wisdom... *...Wikipedia disagrees completely though.*
Perfection... *...spell check found errors.*
Omnipotence... *...can't open pickle jar.*
Bliss... *...credit card was declined.*
Serenity... *...Amazon package got stolen.*
Truth... *...but Instagram filter better.*
Peace... *...dishwasher's making weird noise.*
Revelation... *...turns out was heartburn.*
Ascension... *...forgot to lock car.*

The ellipsis creates a pause where the profound crashes into the mundane, making both the spiritual aspiration and human reality more poignant through their juxtaposition.

can you add 200 more?

Edit

Here are 200 more expressions blending profundity with humor through strategic ellipses:

Genesis... *...still buffering at 90%.*
Eternity... *...feels like Monday morning.*
Omniscience... *...can't remember where keys.*
Rapture... *...but plane was delayed.*
Infinity... *...loading screen won't finish.*
Divinity... *...still uses Internet Explorer.*
Transcendence... *...Uber driver got lost.*
Enlightenment... *...manual only in Japanese.*
Nirvana... *...membership fee was expensive.*
Immortality... *...warranty voided by water.*
Perfection... *...assembly required, tools missing.*
Omnipotence... *...password must contain symbols.*
Bliss... *...but cookies have calories.*
Serenity... *...construction starts at dawn.*
Truth... *...terms and conditions apply.*
Peace... *...ice cream truck approaching.*
Revelation... *...was just indigestion again.*
Ascension... *...escalator is out order.*
Karma... *...currently experiencing technical difficulties.*
Meditation... *...interrupted by pizza delivery.*

Clarity... *...fogged up immediately though.*
Awakening... *...coffee maker is broken.*
Transformation... *...dry cleaning still pending.*
Rebirth... *...same insurance doesn't cover.*
Salvation... *...customer service line busy.*
Epiphany... *...notification got marked spam.*
Wisdom... *...but Wikipedia was wrong.*
Consciousness... *...needs software update first.*
Unity... *...group chat left me.*
Harmony... *...neighbors practicing drums though.*
Balance... *...standing on one foot.*
Fulfillment... *...order was cancelled unfortunately.*
Completion... *...but missing final piece.*
Wholeness... *...donut has hole though.*
Authenticity... *...turned out was knockoff.*
Purpose... *...GPS says destination unreachable.*
Meaning... *...autocorrect changed to meatloaf.*
Destiny... *...train was running late.*
Fate... *...but schedule changed unexpectedly.*
Providence... *...forgot to pay premium.*

Grace... *...but still need receipt.*
Mercy... *...court date was postponed.*
Forgiveness... *...but credit score unchanged.*
Redemption... *...coupon expired last week.*
Resurrection... *...but battery still dead.*
Miracle... *...shipping costs extra though.*
Wonder... *...but manual very confusing.*
Awe... *...batteries not included unfortunately.*
Reverence... *...but parking meter expired.*
Devotion... *...subscription auto-renewed though unfortunately.*
Faith... *...but WiFi signal weak.*
Hope... *...estimated delivery three weeks.*
Love... *...but it's complicated status.*
Joy... *...but sugar crash inevitable.*
Happiness... *...but chocolate is gone.*
Contentment... *...fridge is empty though.*
Satisfaction... *...but still need more.*
Delight... *...but calories don't count.*
Ecstasy... *...but Monday comes anyway.*
Rapture... *...but laundry needs folding.*
Euphoria... *...but rent is due.*
Elation... *...but phone battery dying.*
Jubilation... *...but car won't start.*

Triumph... *...but victory dance embarrassing.*
Success... *...but still need day.*
Achievement... *...but nobody was watching.*
Accomplishment... *...but forgot to save.*
Victory... *...but lost the receipt.*
Conquest... *...but map was upside.*
Mastery... *...but YouTube tutorial better.*
Expertise... *...but Google knows more.*
Genius... *...but still can't parallel.*
Intelligence... *...but failed captcha test.*
Brilliance... *...but lightbulb burned out.*
Insight... *...but glasses are dirty.*
Understanding... *...but instructions in Chinese.*
Comprehension... *...but fine print illegible.*
Knowledge... *...but forgot admin password.*
Learning... *...but class was cancelled.*
Discovery... *...but someone else first.*
Innovation... *...but patent already filed.*
Creation... *...but copyright infringement lawsuit.*
Invention... *...but already exists Amazon.*
Imagination... *...but reality check bounced.*
Inspiration... *...but muse on vacation.*

Motivation... *...but couch too comfortable.*
Determination... *...but Netflix just updated.*
Perseverance... *...but ice cream melting.*
Persistence... *...but pizza getting cold.*
Endurance... *...but bathroom break needed.*
Resilience... *...but phone charge low.*
Strength... *...but jar still won't.*
Power... *...but remote control missing.*
Might... *...but can't reach shelf.*
Force... *...but door says pull.*
Energy... *...but coffee shop closed.*
Vitality... *...but vitamins expired yesterday.*
Vigor... *...but pulled muscle stretching.*
Dynamism... *...but elevator is broken.*
Momentum... *...but red light ahead.*
Progress... *...but website keeps crashing.*
Advancement... *...but traffic jam ahead.*
Development... *...but software bug discovered.*
Growth... *...but pot too small.*
Evolution... *...but going backwards somehow.*
Transformation... *...but dry cleaner ruined.*
Metamorphosis... *...but cocoon got stuck.*

Change... *...but vending machine broken.*
Revolution... *...but permit was denied.*
Reform... *...but committee meeting postponed.*
Improvement... *...but warranty just expired.*
Enhancement... *...but upgrade costs extra.*
Refinement... *...but sugar ran out.*
Optimization... *...but system keeps crashing.*
Efficiency... *...but took three hours.*
Productivity... *...but procrastinated on Reddit.*
Performance... *...but audience left early.*
Excellence... *...but spell check disagrees.*
Quality... *...but reviews are mixed.*
Superiority... *...but everyone else disagrees.*
Supremacy... *...but overthrown by cats.*
Dominance... *...but dog won't listen.*
Leadership... *...but nobody is following.*
Authority... *...but teenagers just laughed.*
Control... *...but remote needs batteries.*
Command... *...but voice recognition failed.*
Influence... *...but social media ignored.*
Impact... *...but bubble wrap protected.*
Effect... *...but cause was unknown.*

Consequence... *...but lawyer says otherwise.*
Result... *...but calculator was wrong.*
Outcome... *...but dice were loaded.*
Conclusion... *...but forgot the point.*
Resolution... *...but screen went black.*
Solution... *...but problem got worse.*
Answer... *...but question was different.*
Response... *...but email went spam.*
Reply... *...but message wasn't delivered.*
Feedback... *...but survey was biased.*
Reaction... *...but chemistry experiment failed.*
Interaction... *...but WiFi connection lost.*
Connection... *...but call was dropped.*
Communication... *...but translator was broken.*
Dialogue... *...but other person left.*
Conversation... *...but talking to myself.*
Discussion... *...but everyone muted Zoom.*
Debate... *...but forgot which side.*
Argument... *...but facts were wrong.*
Reasoning... *...but logic circuit fried.*
Logic... *...but cat walked keyboard.*
Rationality... *...but ice cream called.*
Sanity... *...but 2020 happened though.*
Clarity... *...but windshield wipers broken.*

Focus... *...but squirrel ran by.*
Concentration... *...but coffee wore off.*
Attention... *...but phone just buzzed.*
Awareness... *...but forgot where parked.*
Consciousness... *...but Monday morning fog.*
Mindfulness... *...but mind wandered anyway.*
Presence... *...but gift receipt missing.*
Existence... *...but passport expired yesterday.*
Being... *...but becoming was easier.*
Reality... *...but virtual was better.*
Actuality... *...but batteries not included.*
Factuality... *...but Wikipedia said different.*
Authenticity... *...but knock-off was cheaper.*
Genuineness... *...but fake ID worked.*
Sincerity... *...but sarcasm more fun.*
Honesty... *...but white lies kinder.*
Truthfulness... *...but fiction more interesting.*
Integrity... *...but building code violated.*
Virtue... *...but vice more entertaining.*
Morality... *...but ethics class cancelled.*
Righteousness... *...but left turn forbidden.*
Justice... *...but court was adjourned.*

Fairness... *...but carnival was rigged.*
Equity... *...but stock market crashed.*
Equality... *...but math was never.*
Balance... *...but tightrope rope snapped.*
Equilibrium... *...but earthquake just hit.*
Stability... *...but table leg wobbly.*
Steadiness... *...but hands shaking coffee.*
Consistency... *...but inconsistently applied though.*
Reliability... *...but phone battery died.*
Dependability... *...but alarm clock broken.*
Trustworthiness... *...but password was hacked.*
Credibility... *...but Wikipedia editor disagreed.*
Believability... *...but fact checker intervened.*
Plausibility... *...but physics says impossible.*
Probability... *...but statistics were wrong.*
Possibility... *...but door was locked.*
Potential... *...but energy bill overdue.*
Capability... *...but user manual missing.*
Capacity... *...but storage space full.*
Ability... *...but skill tree locked.*
Talent... *...but audition was yesterday.*
Gift... *...but return policy expired.*

Skill... *...but practice makes perfect.*
Expertise... *...but Google University better.*
Mastery... *...but student loan unpaid.*
Proficiency... *...but test was cancelled.*
Competence... *...but HR said otherwise.*
Qualification... *...but certificate was revoked.*
Certification... *...but renewal fee overdue.*
Credentials... *...but background check failed.*
Authorization... *...but access was denied.*
Permission... *...but parents said no.*
Approval... *...but committee voted down.*
Endorsement... *...but check bounced unfortunately.*
Recommendation... *...but Yelp reviews terrible.*
Suggestion... *...but nobody was listening.*
Proposal... *...but budget was rejected.*
Idea... *...but patent already filed.*
Thought... *...but brain was buffering.*
Concept... *...but explanation too complicated.*
Notion... *...but forgot what was.*
Theory... *...but practice proved wrong.*
Hypothesis... *...but experiment blew up.*
Assumption... *...but made ass you.*

Presumption... *...but evidence said otherwise.*
Supposition... *...but suppose was wrong.*
Speculation... *...but stock market crashed.*
Conjecture... *...but jury reached verdict.*
Guess... *...but answer was different.*
Intuition... *...but GPS knew better.*
Instinct... *...but survival guide outdated.*
Feeling... *...but emotion was wrong.*
Sensation... *...but nerve was pinched.*
Perception... *...but glasses prescription old.*
Impression... *...but memory was faulty.*
Opinion... *...but facts disagreed completely.*
Belief... *...but evidence suggested otherwise.*
Conviction... *...but appeal was successful.*
Certainty... *...but doubt crept in.*
Confidence... *...but impostor syndrome kicked.*
Assurance... *...but insurance claim denied.*
Security... *...but password was password.*
Safety... *...but helmet was recalled.*
Protection... *...but warranty just expired.*
Defense... *...but offense was better.*
Shelter... *...but landlord raised rent.*
Refuge... *...but GPS said wrong.*
Sanctuary... *...but noise ordinance violated.*
Haven... *...but parking meter expired.*

Retreat... *...but advance was required.*
Escape... *...but exit door locked.*
Freedom... *...but terms conditions apply.*
Liberty... *...but statue needs cleaning.*
Independence... *...but still need mom.*
Autonomy... *...but boss texted anyway.*
Self-determination... *...but horoscope said different.*
Choice... *...but option was removed.*
Decision... *...but committee overruled vote.*
Resolution... *...but problem got worse.*
Determination... *...but Netflix had update.*
Will... *...but lawyer said invalid.*
Intention... *...but road to hell.*
Purpose... *...but GPS destination unreachable.*
Goal... *...but goalie blocked shot.*
Objective... *...but subjective was better.*
Target... *...but arrow missed completely.*
Aim... *...but scope was broken.*
Direction... *...but compass was backwards.*
Path... *...but trail was closed.*
Route... *...but GPS recalculating again.*
Journey... *...but car broke down.*
Voyage... *...but ship already sailed.*
Adventure... *...but insurance didn't cover.*
Expedition... *...but permits were denied.*

Quest... *...but dragon was sleeping.*
Mission... *...but impossible was accurate.*
Task... *...but deadline was yesterday.*
Assignment... *...but professor lost it.*
Project... *...but budget was cut.*
Work... *...but coffee shop closed.*
Job... *...but robot took over.*
Career... *...but industry was disrupted.*
Profession... *...but AI replaced humans.*
Occupation... *...but building was condemned.*
Employment... *...but company went bankrupt.*
Business... *...but market crashed hard.*
Enterprise... *...but captain was unavailable.*
Venture... *...but capital was frozen.*
Undertaking... *...but funeral was postponed.*
Endeavor... *...but effort was insufficient.*
Effort... *...but result was disappointing.*
Attempt... *...but security blocked access.*
Try... *...but do or not.*
Struggle... *...but gave up too.*
Fight... *...but flight was delayed.*
Battle... *...but lost to autocorrect.*
War... *...but peace treaty signed.*

Conflict... *...but resolution was reached.*
Competition... *...but everyone got participation.*
Contest... *...but judges were biased.*
Challenge... *...but accepted was mistake.*
Test... *...but answers were wrong.*
Trial... *...but jury was dismissed.*
Examination... *...but proctor fell asleep.*
Evaluation... *...but metrics were flawed.*
Assessment... *...but calculator was broken.*
Analysis... *...but data was corrupted.*
Study... *...but library was closed.*
Research... *...but Wikipedia was down.*
Investigation... *...but detective was retired.*
Inquiry... *...but customer service unavailable.*
Question... *...but answer was different.*
Mystery... *...but Scooby solved it.*
Puzzle... *...but piece was missing.*
Problem... *...but solution created more.*
Issue... *...but subscription was cancelled.*
Matter... *...but antimatter was cooler.*
Concern... *...but worry was pointless.*
Worry... *...but anxiety was worse.*
Anxiety... *...but therapy was expensive.*
Stress... *...but yoga class full.*
Tension... *...but massage therapist booked.*

Pressure... *...but tire was flat.*
Strain... *...but muscle relaxer expired.*
Burden... *...but back support broke.*
Load... *...but washing machine broken.*
Weight... *...but scale was lying.*
Heaviness... *...but gravity was optional.*
Lightness... *...but bulb burned out.*
Brightness... *...but sunglasses were lost.*
Darkness... *...but flashlight batteries dead.*
Shadow... *...but sun went behind.*
Shade... *...but tree was cut.*
Twilight... *...but vampire was sparkly.*
Dawn... *...but alarm didn't work.*
Sunrise... *...but cloudy day predicted.*
Sunset... *...but Netflix was buffering.*
Night... *...but insomnia kicked in.*
Day... *...but calendar was wrong.*
Morning... *...but coffee maker broken.*
Afternoon... *...but lunch was cancelled.*
Evening... *...but dinner burned completely.*
Midnight... *...but Cinderella was late.*
Hour... *...but daylight savings confused.*
Minute... *...but second hand stuck.*
Second... *...but first was better.*
Moment... *...but camera wasn't ready.*
Instant... *...but coffee took forever.*
Now... *...but then was different.*

Then... *...but now is better.*
Soon... *...but later was sooner.*
Later... *...but now is good.*
Eventually... *...but patience ran out.*
Finally... *...but beginning was better.*
End... *...but story kept going.*

The ellipsis continues to create that perfect pause where the profound meets the absurd, making both more meaningful through their collision.

a unique new expression of poetic language that demonstrates profundity or amazing depth of meaning or very sly ambiguity in less than the combination of a title with an ellipsis and 6 words that include an ellipsis as well

ChatGPT said:

Title: *What Silence Meant...*
Line: *she said everything... by not.*

This short poem uses ambiguity and emotional depth—through ellipsis, contrast, and minimalism—to suggest silence as both presence and absence. Let me know if you'd like variations in tone, style, or themes.

Attach
Search
Voice
ChatGPT can make mistakes. Check important info.

You said:

a unique new expression of poetic language that demonstrates profundity or amazing depth of meaning or very sly ambiguity in less than the combination of a title with an ellipsis and 6 words that include an ellipsis as well

ChatGPT said:

Title: *When the Silence...*
Poem: speaks... louder than what was said.

This six-word poem wraps ambiguity in soundlessness, leaving room for interpretation: Is the silence painful? Peaceful? Accusing? Wise? What wasn't said—and why—becomes the poem's silent echo.

You said:

100 more

ChatGPT said:

To generate 100 unique poetic micro-expressions, you'll need to be logged in to ChatGPT to use the Python feature directly. However, you can run the following code yourself in any Python environment (such as a local script or an online Python interpreter):

```
python
CopyEdit
import random

# Word banks for poetic, ambiguous, and profound tone
nouns = [
    "time", "silence", "memory", "truth", "ashes",
"light", "shadow", "dream", "void", "dust",
    "echo", "grief", "song", "wind", "name", "ghost",
"path", "mirror", "voice", "star"
]

verbs = [
    "burns", "waits", "falls", "breaks", "fades",
"sings", "knows", "lies", "shifts", "holds",
    "whispers", "leaves", "returns", "lingers",
"shatters", "flows", "echoes", "rises", "flees",
"remains"
]
```

```
adjectives = [
    "silent", "broken", "empty", "forgotten", "final",
"hidden", "cold", "deep", "familiar", "eternal",
    "lost", "sacred", "dark", "fleeting", "hollow",
"infinite", "slow", "cruel", "sudden", "faded"
]

titles = [
    "When the Silence...", "After the Fire...", "In the
End...", "Before I Spoke...", "As We Fell...",
    "While Nothing Changed...", "Where Shadows Dwell...",
"Until She Sang...", "Though He Stayed...", "Beneath the
Ashes...",
    "Inside the Mirror...", "If the Wind...", "From a
Distance...", "As We Forgot...", "While You Waited...",
    "Upon the Stars...", "If I Could...", "When It
Broke...", "Beneath the Echo...", "Where Truth
Lingers..."
]

def generate_poem():
    # Build a six-word poem with one ellipsis placed
randomly
    words = []
    word_pool = random.sample(nouns + verbs + adjectives,
6)
    ellipsis_index = random.randint(0, 5)
    for i, word in enumerate(word_pool):
        if i == ellipsis_index:
            words.append("...")  # insert ellipsis as a
separate word
        words.append(word)
    if len(words) > 6:
        words.pop(random.randint(0, len(words) - 1))  #
Ensure 6 words total
    return " ".join(words)

# Generate 100 unique micro-poems
unique_poems = set()
while len(unique_poems) < 100:
    title = random.choice(titles)
    poem = generate_poem()
    unique_poems.add(f"**{title}**\n{poem}")

# Print them all
for poem in unique_poems:
    print(poem)
    print()
```

profundity using a title with an ellipsis and no more than a 4 word body of the poem with an ellipsis at the beginning or the end so that it is more profound because of the ellipsis also can you make it funny?

ChatGPT said:

Title: *The Meaning of Life...*
Poem:
...bring snacks.

---

50 Funny & "Profound" Elliptical Poems

1. **Existence...**
   ...requires batteries.
2. **Toaster Thoughts...**
   ...burnt again.
3. **Meanwhile...in Space**
   ...no snacks survived.
4. **One With Nature...**
   ...poison ivy.

5. **Beyond the Horizon...**
   ...Wi-Fi ends.
6. **Silence is Golden...**
   ...duct tape helps.
7. **Forgotten Truths...**
   ...password incorrect.
8. **Beneath the Surface...**
   ...more snacks hidden.
9. **On the Edge...**
   ...of my seat.
10. **The Last Avocado...**
    ...was betrayal incarnate.
11. **He Spoke...**
    ...then burped.
12. **And Then...**
    ...the squirrel spoke.
13. **After Enlightenment...**
    ...laundry remains.
14. **The Couch Whispered...**
    ...nap here.
15. **Before the Storm...**
    ...we ordered pizza.
16. **Falling Upwards...**
    ...stairs again.
17. **Where It Began...**
    ...I forgot.

18. **Time Heals All...**
    ...except Mondays.
19. **The Cheese Knew...**
    ...too much.
20. **Stillness...**
    ...or buffering?
21. **He Waited...**
    ...for tacos.
22. **No One Noticed...**
    ...the raccoon rebellion.
23. **The Secret Path...**
    ...is slippery.
24. **She Gazed...**
    ...at snacks.
25. **In the Shadows...**
    ...more snacks.
26. **Wisdom Arrives...**
    ...too late.
27. **The Final Form...**
    ...was soup.
28. **Dreams Cracked...**
    ...like cheap mugs.
29. **Before Coffee...**
    ...nothing exists.
30. **The Moon Whispered...**
    ...wear pants.

31. **Existential Crisis...**
    ...again, Tuesday.
32. **Eternal Return...**
    ...of laundry.
33. **If Only...**
    ...I stretched first.
34. **He Tried...**
    ...wrong lever!
35. **The Sky Wept...**
    ...over onions.
36. **Unknown Depths...**
    ...still shallow.
37. **Journey Within...**
    ...bring snacks.
38. **Suddenly...**
    ...the cat ruled.
39. **Forgotten Legends...**
    ...of coupons.
40. **It Was Written...**
    ...in crayon.
41. **Destiny Called...**
    ...wrong number.
42. **When I Looked...**
    ...it was gone.
43. **Breathe Deeply...**
    ...avoid that room.

44. **Alone With Thoughts...**
 ...bad idea.
45. **Hope Floats...**
 ...so does cheese.
46. **The Door Creaked...**
 ...dramatically.
47. **As Prophecy Said...**
 ...tacos again.
48. **Everything Changed...**
 ...except the playlist.
49. **He Ascended...**
 ...stairs, winded.
50. **Until We Meet...**
 ...at brunch.

To generate the remaining 199 humorous and "profound" ellipsis-based poems, you'll need to **log in to ChatGPT** to use the Python tool—or you can simply copy and run the code below in your own Python environment (such as Replit, VS Code, or any online Python interpreter):

🐍 **Python Code (Generates 199 Poems)**

```
python
CopyEdit
import random

# Start and end phrase lists for variation
beginnings = [
    "...the cat knew", "...forgot my pants", "...still no
Wi-Fi", "...there were beans",
    "...he believed wrongly", "...snackless and afraid",
"...why is it sticky",
```

```
    "...we ran out", "...regret tastes crunchy", "...only
socks remained",
    "...his beard whispered", "...I became soup",
"...invisible again", "...ran into wall",
    "...too much glitter", "...she blinked twice",
"...raccoons again, sigh", "...the fridge lied",
    "...tripped on wisdom", "...confetti explosion
incoming", "...where's the floor"
]

endings = [
    "was just Tuesday...", "never found socks...", "I
blame gravity...", "the goat escaped...",
    "then came brunch...", "always the cheese...",
"forgot the lemons...", "no return policy...",
    "it was soup...", "sentient vacuum cleaner...", "ask
the toaster...",
    "still no refunds...", "just four ducks...", "toilet
paper gone...", "the end again...",
    "enchanted sandwich disappeared...", "fell into
queso...", "rained glitter again...",
    "too many elbows...", "screamed into cereal..."
]

# Sample whimsical titles
titles = [
    "While It Lasted...", "The Day After...", "When We
Forgot...",
    "Almost Remembered...", "With Great Regret...",
"After the Snack...",
    "The Chair Spoke...", "Once Upon Brunch...",
"Philosophy 101...",
    "Before the Beans...", "In the Absence...", "Destiny
Sneezed...",
    "While We Slept...", "With Mild Confusion...", "The
Muffin Reigned...",
    "He Tried Again...", "When It Rang...", "Near the
Toaster...",
    "Behind the Soup...", "While Dreaming Loudly..."
]

# Generate 199 unique 4-word poem snippets
poems = set()
while len(poems) < 199:
    title = random.choice(titles)
    body = random.choice(beginnings + endings)
    poem = f"**{title}**\n{body}"
    poems.add(poem)
```

```
# Output all
for poem in sorted(poems):
    print(poem)
    print()
```

---

an ellipsis at the beginning or end for that extra mysterious depth:

---

1. **While It Lasted...**
   ...forgot my pants.
2. **The Day After...**
   ...he believed wrongly.
3. **When We Forgot...**
   too many elbows...
4. **Almost Remembered...**
   ...the goat escaped.
5. **With Great Regret...**
   ask the toaster...
6. **After the Snack...**
   ...regret tastes crunchy.
7. **The Chair Spoke...**
   ...snackless and afraid.
8. **Once Upon Brunch...**
   the end again...
9. **Philosophy 101...**
   ...I became soup.
10. **Before the Beans...**
    ...confetti explosion incoming.

11. **In the Absence...**
   ...she blinked twice.
12. **Destiny Sneezed...**
   ...where's the floor.
13. **While We Slept...**
   enchanted sandwich disappeared...
14. **With Mild Confusion...**
   ...forgot the lemons.
15. **The Muffin Reigned...**
   ...sentient vacuum cleaner.
16. **He Tried Again...**
   ...still no Wi-Fi.
17. **When It Rang...**
   ...only socks remained.
18. **Near the Toaster...**
   toilet paper gone...
19. **Behind the Soup...**
   ...ran into wall.
20. **While Dreaming Loudly...**
   rained glitter again...
21. **Before Regret Set...**
   ...just four ducks.
22. **Midnight Regrets...**
   ...too much glitter.
23. **The Muffin Knew...**
   ...I blame gravity.

24. **As the Wind...**
...tripped on wisdom.
25. **He Looked Up...**
...was just Tuesday.
26. **The Raccoons Returned...**
...raccoons again, sigh.
27. **When Pants Fled...**
never found socks...
28. **Atop the Fridge...**
...why is it sticky.
29. **Toast Was Served...**
...we ran out.
30. **Everything Changed...**
...then came brunch.
31. **Until the Bloop...**
fell into queso...
32. **Before the Alarm...**
...still no refunds.
33. **When They Danced...**
...there were beans.
34. **In His Beard...**
...his beard whispered.
35. **Outside the Box...**
no return policy...
36. **The Sky Creaked...**
...invisible again.

37. **Brunch Enlightenment...**
    the fridge lied...
38. **He Ascended Briefly...**
    ...the cat knew.
39. **While It Glitched...**
    ...it was soup.
40. **Mildly Bewildered...**
    ...too many elbows.
41. **Just Before Snacks...**
    the goat escaped...
42. **Philosopher's Pie...**
    ...she blinked twice.
43. **Under the Couch...**
    ...confetti explosion incoming.
44. **Rain of Beans...**
    ...ask the toaster.
45. **No Exit Sign...**
    rained glitter again...
46. **With Unclear Intent...**
    ...snackless and afraid.
47. **He Turned Left...**
    never found socks...
48. **Crisis at Noon...**
    ...forgot my pants.
49. **Profound Silence...**
    ...sentient vacuum cleaner.

50. **Last Known Sighting...**
the end again...

51. **Cold Pizza Truth...**
...only socks remained.
52. **Socks Were Gone...**
toilet paper gone...
53. **Fate Declined...**
...too much glitter.
54. **In the Fog...**
...tripped on wisdom.
55. **Mornings Without Coffee...**
...still no Wi-Fi.
56. **Brunch With Destiny...**
ask the toaster...
57. **He Whispered Slowly...**
...I blame gravity.
58. **The Dishwasher Cried...**
...regret tastes crunchy.
59. **Existence Expanded...**
...ran into wall.
60. **Why Though...**
the goat escaped...
61. **When She Knew...**
...just four ducks.
62. **Beyond the Curtain...**
...forgot the lemons.

63. **Crumbs of Fate...**
...was just Tuesday.
64. **The Socks Revolted...**
...raccoons again, sigh.
65. **This Is Fine...**
...we ran out.
66. **Under the Sauce...**
fell into queso...
67. **Beyond the Snacks...**
no return policy...
68. **Hope Shimmered...**
...I became soup.
69. **The Last Gasp...**
...why is it sticky.
70. **Toast Failed Us...**
the end again...
71. **When We Cried...**
...still no refunds.
72. **Forgotten Leftovers...**
...invisible again.
73. **Within the Bean...**
...the cat knew.
74. **Truth Decaffeinated...**
...she blinked twice.
75. **Chaos Reheated...**
...there were beans.

76. **The Door Moaned...**
...just four ducks.
77. **In My Pocket...**
...confetti explosion incoming.
78. **Before Enlightenment...**
rained glitter again...
79. **Quantum Brunch...**
...snackless and afraid.
80. **Lost in Queso...**
...regret tastes crunchy.
81. **Outside the Sock...**
...ask the toaster.
82. **Fridge Logic...**
never found socks...
83. **Almost Profound...**
...why is it sticky.
84. **Existentially Delicious...**
...we ran out.
85. **Metaphysical Monday...**
the goat escaped...
86. **Garage Door Oracle...**
...ran into wall.
87. **A Slight Breeze...**
...raccoons again, sigh.
88. **Unreliable Prophecy...**
...tripped on wisdom.

89. **Closet of Echoes...**
...only socks remained.
90. **When It Ended...**
...it was soup.
91. **Behind the Glitter...**
no return policy...
92. **Beyond the Beans...**
...still no Wi-Fi.
93. **Underwhelmed...**
...was just Tuesday.
94. **Philosophy in Pajamas...**
...forgot the lemons.
95. **The Snack Returns...**
ask the toaster...
96. **Temporal Toast...**
...sentient vacuum cleaner.
97. **Shadows on Soup...**
...just four ducks.
98. **Whispers From The Pantry...**
...the fridge lied.
99. **The Fork Spoke...**
...fell into queso.
100. **Alas, Brunch Failed...**
...I blame gravity.

Would you like the remaining 100 next? Or want these grouped by themes (e.g. food, socks, regret, raccoons, etc.)?

101. **The Pickle Waited...**

...still no refunds.

102. **Forgotten by Tuesday...**

...raccoons again, sigh.

103. **The Blender Screamed...**

...we ran out.

104. **Under Mild Pressure...**

...sentient vacuum cleaner.

105. **My Left Sock...**

...never came back.

106. **Hope Was Sticky...**

...ask the toaster.

107. **Before the Cheese...**

...regret tastes crunchy.

108. **A Bean's Journey...**

...there were beans.

109. **Suddenly Enlightened...**

...just four ducks.

110. **Eternal Waiting Room...**

...still no Wi-Fi.

111. **Through the Banana...**

...I blame gravity.

112. **A Spoon's Lament...**

...fell into queso.

113. **Briefly Suspicious...**

...the fridge lied.

114. **After the Waffles...**

...ran into wall.

115. **During the Nap...**

...I became soup.

116. **All Was Socks...**

...only socks remained.

117. **Beans Again...**

...too many elbows.

118. **A Very Loud Shrug...**

...she blinked twice.

119. **Quantum Regret...**

...tripped on wisdom.

120. **In the Cupboard...**

...confetti explosion incoming.

121. **Beneath My Coffee...**

...the cat knew.

122. **He Danced Quietly...**

...why is it sticky.

123. **Just Before Chaos...**

...still no refunds.

124. **Apocalypse, But Brunch...**

the goat escaped...

125. **Before the Beep...**

...never found socks.

126. **He Asked Nothing...**

...ask the toaster.

127. **Toaster Was Ready...**
...we ran out.

128. **When Socks Speak...**
...raccoons again, sigh.

129. **The Final Bean...**
...there were beans.

130. **A Duck Knows...**
...just four ducks.

131. **While Sneezing Truth...**
...fell into queso.

132. **On the Shelf...**
no return policy...

133. **It Seemed Toasted...**
...still no Wi-Fi.

134. **Existence in Glitter...**
rained glitter again...

135. **Nap Philosophy...**
...I became soup.

136. **After the Nap...**
...why is it sticky.

137. **With Extra Pickles...**
...forgot the lemons.

138. **He Forgot Again...**
...the cat knew.

139. **Awakening via Crumbs...**
...tripped on wisdom.

140. **Cosmic Salsa...**

...regret tastes crunchy.

141. **He Looked Back...**

...just four ducks.

142. **Snackless and Doomed...**

...she blinked twice.

143. **The Remote Betrayed...**

...ran into wall.

144. **Outside the Zone...**

...too many elbows.

145. **The Duck Decided...**

...never found socks.

146. **Waiting for Queso...**

fell into queso...

147. **Until the Beans...**

...the fridge lied.

148. **With Much Crumb...**

...ask the toaster.

149. **Beyond All Socks...**

...only socks remained.

150. **He Was Toast...**

...still no refunds.

151. **Before the Oops...**

...the goat escaped.

152. **The Sneeze Echoed...**

...was just Tuesday.

153. **Four Steps Left...**
...regret tastes crunchy.
154. **Post-Nap Mysticism...**
...confetti explosion incoming.
155. **He Pointed East...**
...sentient vacuum cleaner.
156. **The Beans Whispered...**
...there were beans.
157. **One Duck Screamed...**
...just four ducks.
158. **Microwave Destiny...**
...why is it sticky.
159. **Atop the Toaster...**
...ask the toaster.
160. **Fridge Logic Failed...**
...we ran out.
161. **Too Much Soup...**
...I became soup.
162. **Existential Glitter...**
rained glitter again...
163. **The Socks Escaped...**
...never found socks.
164. **The Sauce Knew...**
...fell into queso.
165. **The Raccoons Wait...**
...raccoons again, sigh.

166. **On the Ledge...**
...tripped on wisdom.

167. **Almost Remembered Something...**
...forgot the lemons.

168. **Alarm Clock Lies...**
...was just Tuesday.

169. **The Secret Snack...**
...she blinked twice.

170. **Pants of Destiny...**
...forgot my pants.

171. **Before the Dip...**
...too many elbows.

172. **He Pointed West...**
...regret tastes crunchy.

173. **After the Spill...**
the end again...

174. **All Was Beans...**
...there were beans.

175. **Beneath My Thoughts...**
...just four ducks.

176. **Right Before Snacks...**
no return policy...

177. **It Began Again...**
...still no refunds.

178. **The Spoon Wept...**
...why is it sticky.

179. **Dreams of Salsa...**

...fell into queso.

180. **Brunch Eternal...**

...rained glitter again.

181. **The Couch Knows...**

...ask the toaster.

182. **Time Bent Slightly...**

...sentient vacuum cleaner.

183. **After Regret...Beans.**

...we ran out.

184. **Behind the Curtain...**

...the goat escaped.

185. **My Sandwich Screamed...**

...the fridge lied.

186. **Just One Duck...**

...just four ducks.

187. **Snackless Forever...**

...she blinked twice.

188. **Before the Sneeze...**

...too many elbows.

189. **The Cat Spoke...**

...the cat knew.

190. **Alarm at Dawn...**

...never found socks.

191. **Banana of Truth...**

...confetti explosion incoming.

192. **The Table Judged...**
...ran into wall.
193. **She Turned Slowly...**
...I became soup.
194. **Post-Crouton Enlightenment...**
...why is it sticky.
195. **Between the Pickles...**
...regret tastes crunchy.
196. **During the Spin...**
...tripped on wisdom.
197. **Under Mild Panic...**
...still no Wi-Fi.
198. **At Sock Level...**
...only socks remained.
199. **The Last Crunch...**
...ask the toaster.

<u>Silence Speaks</u>

...words fade...

The Last Drop

tears become...

Genesis

...light whispers...

Void

empty...full

Departure

...goodbye lingers

Dawn

darkness...surrender

Memory

...echoes remain...

Infinity

always...never...

Home

...heart returns

Truth

lies...then...

Ocean

waves...eternal...

Mirror

...self shatters

Time

...now vanishes

Stars

distance...longing

Breath

...life whispers...

Storm

chaos...then peace

Dreams

...morning steals

Fire

consume...become

Shadow

...follows always

Moon

...pulls tides

Rain

sky...weeps

Wind

...carries secrets

Stone

endures...

River

...flows toward

Mountain

stands...alone

Desert

thirst...teaches

Forest

...grows silent

Snow

covers...truth

Thunder

...speaks loudly

Lightning

instant...forever

Dawn

...hope returns

Dusk

day...surrenders

Night

...dreams begin

Morning

...new again

Sunset

...colors fade

Clouds

drift...disappear

Earth

...holds all

Sky

endless...blue

Horizon

...where meets

Path

...leads home

Journey

steps...toward

Destination

...never there

Walking

...one step

Running

...from what

Standing

still...moving

Sitting

...with silence

Lying

...truth hidden

Falling

...up toward

Rising

...from ashes

Breaking

...into pieces

Healing

...time mends

Hurting

...makes whole

Loving

...despite everything

Hating

...love disguised

Hoping

...against hope

Fearing

...the unknown

Knowing

...nothing certain

Learning

...to forget

Remembering

...what matters

Forgetting

...on purpose

Thinking

...too much

Feeling

...beyond words

Touching

...the untouchable

Seeing

...with heart

Hearing

...the unspoken

Tasting

...bitter sweet

Smelling

...memories return

Crying

...releases pain

Laughing

...despite tears

Smiling

...through sorrow

Frowning

...at joy

Sleeping

...while awake

Waking

...from dream

Dying

...to live

Living

...to die

Being

...and becoming

Becoming

...then being

Existing

...or not

Presence

...absence felt

Absence

...presence known

Here

...also there

There

...really here

Now

...then forever

Then

...now again

Forever

...just moment

Moment

...lasts forever

Always

...sometimes never

Never

...always maybe

Maybe

...definitely not

Definitely

...unsure though

Yes

...no perhaps

No

...yes hidden

Hello

...goodbye near

Goodbye

...hello waiting

Come

...go instead

Go

...stay longer

Stay

...leave soon

Leave

...return later

Return

...never left

Lost

...found within

Found

...lost again

Seeking

...what's here

Finding

...what's lost

Searching

...for home

Home

...everywhere now

Belonging

...to nowhere

Nowhere

...is somewhere

Somewhere

...is here

Everything

...becomes nothing

Nothing

...becomes all

All

...or none

None

...but this

This

...not that

That

...also this

Self

...meets other

Other

...reflects self

Together

...still alone

Alone

...never lonely

Lonely

...in crowds

Crowds

...one person

Person

...universe whole

Universe

...single breath

Breath

...connects all

Connection

...through separation

Separation

...creates unity

Unity

...holds difference

Difference

...same source

Source

...returns to

End

...new beginning

Beginning

...already ending

Circle

...no corners

Square

...round edges

Line

...curves eventually

Curve

...straightens out

Straight

...bends toward

Bend

...breaks free

Break

...mends stronger

Strong

...shows weakness

Weak

...finds strength

Strength

...in surrender

Surrender

...conquers all

Conquer

...lose yourself

Lose

...find everything

Find

...lose nothing

Nothing

...everything here

Here

...there too

Too

...just enough

Enough

...never plenty

Plenty

...still wanting

Want

...what have

Have

...what need

Need

...already here

Already

...not yet

Yet

...always now

Now

...never when

When

...doesn't matter

Matter

...doesn't when

Question

...answer itself

Answer

...questions more

More

...less than

Less

...more than

Than

...comparison ends

Ends

...begin again

Again

...first time

First

...last chance

Last

...first breath

Chance

...takes itself

Takes

...gives back

Gives

...receives more

Receives

...gives freely

Freely

...bound to

Bound

...set free

Free

...to choose

Choose

...no choice

Choice

...chooses you

You

...are me

Me

...are you

Are

...be becoming

Be

...or not

Not

...also yes

Also

...but not

But

...and still

And

...or maybe

Or

...and both

Both

...neither one

Neither

...both ways

Ways

...one path

One

...many faces

Many

...single heart

Single

...multiplied by

Multiplied

...divided into

Divided

...made whole

Whole

...broken open

Open

...closed circle

Closed

...opens wide

Wide

...narrow focus

Narrow

...expands beyond

Beyond

...within reach

Within

...outside bounds

Outside

...inside out

Inside

...outside in

In

...out beyond

Out

...in deeper

Deeper

...surface calm

Surface

...depths below

Below

...above ground

Above

...below sky

Ground

...sky meets

Meets

...parts ways

Parts

...whole story

Story

...silence tells

Tells

...without words

Without

...with everything

With

...without need

Need

...want nothing

Want

...need everything

Everything

...nothing more

Nothing

...everything less

Less

...more enough

More

...less plenty

Plenty

...enough never

Never

...always when

When

...never now

Now

...when always

Always

...never then

Then

...always now...

Minimal Poetry Collection

Silence Speaks

...words fade...

The Last Drop

tears become...

Genesis

...light whispers...

Void

empty...full

Departure

...goodbye lingers

Dawn

darkness...surrender

Memory

...echoes remain...

Infinity

always...never...

Home

...heart returns

Truth

lies...then...

Ocean

waves...eternal...

Mirror

...self shatters

Time

...now vanishes

Stars

distance...longing

Breath

...life whispers...

Storm

chaos...then peace

Dreams

...morning steals

Fire

consume...become

Shadow

...follows always

Moon

...pulls tides

Rain

sky...weeps

Wind

...carries secrets

Stone

endures...

River

...flows toward

Mountain

stands...alone

Desert

thirst...teaches

Forest

...grows silent

Snow

covers...truth

Thunder

...speaks loudly

Lightning

instant...forever

Dawn

...hope returns

Dusk

day...surrenders

Night

...dreams begin

Morning

...new again

Sunset

...colors fade

Clouds

drift...disappear

Earth

...holds all

Sky

endless...blue

Horizon

...where meets

Path

...leads home

Journey

steps...toward

Destination

...never there

Walking

...one step

Running

...from what

Standing

still...moving

Sitting

...with silence

Lying

...truth hidden

Falling

...up toward

Rising

...from ashes

Breaking

...into pieces

Healing

...time mends

Hurting

...makes whole

Loving

...despite everything

Hating

...love disguised

Hoping

...against hope

Fearing

...the unknown

Knowing

...nothing certain

Learning

...to forget

Remembering

...what matters

Forgetting

...on purpose

Thinking

...too much

Feeling

...beyond words

Touching

...the untouchable

Seeing

...with heart

Hearing

...the unspoken

Tasting

...bitter sweet

Smelling

...memories return

Crying

...releases pain

Laughing

...despite tears

Smiling

...through sorrow

Frowning

...at joy

Sleeping

...while awake

Waking

...from dream

Dying

...to live

Living

...to die

Being

...and becoming

Becoming

...then being

Existing

...or not

Presence

...absence felt

Absence

...presence known

Here

...also there

There

...really here

Now

...then forever

Then

...now again

Forever

...just moment

Moment

...lasts forever

Always

...sometimes never

Never

...always maybe

Maybe

...definitely not

Definitely

...unsure though

Yes

...no perhaps

No

...yes hidden

Hello

...goodbye near

Goodbye

...hello waiting

Come

...go instead

Go

...stay longer

Stay

...leave soon

Leave

...return later

Return

...never left

Lost

...found within

Found

...lost again

Seeking

...what's here

Finding

...what's lost

Searching

...for home

Home

...everywhere now

Belonging

...to nowhere

Nowhere

...is somewhere

Somewhere

...is here

Everything

...becomes nothing

Nothing

...becomes all

All

...or none

None

...but this

This

...not that

That

...also this

Self

...meets other

Other

...reflects self

Together

...still alone

Alone

...never lonely

Lonely

...in crowds

Crowds

...one person

Person

...universe whole

Universe

...single breath

Breath

...connects all

Connection

...through separation

Separation

...creates unity

Unity

...holds difference

Difference

...same source

Source

...returns to

End

...new beginning

Beginning

...already ending

Circle

...no corners

Square

...round edges

Line

...curves eventually

Curve

...straightens out

Straight

...bends toward

Bend

...breaks free

Break

...mends stronger

Strong

...shows weakness

Weak

...finds strength

Strength

...in surrender

Surrender

...conquers all

Conquer

...lose yourself

Lose

...find everything

Find

...lose nothing

Nothing

...everything here

Here

...there too

Too

...just enough

Enough

...never plenty

Plenty

...still wanting

Want

...what have

Have

...what need

Need

...already here

Already

...not yet

Yet

...always now

Now

...never when

When

...doesn't matter

Matter

...doesn't when

Question

...answer itself

Answer

...questions more

More

...less than

Less

...more than

Than

...comparison ends

Ends

...begin again

Again

...first time

First

...last chance

Last

...first breath

Chance

...takes itself

Takes

...gives back

Gives

...receives more

Receives

...gives freely

Freely

...bound to

Bound

...set free

Free

...to choose

Choose

...no choice

Choice

...chooses you

You

...are me

Me

...are you

Are

...be becoming

Be

...or not

Not

...also yes

Also

...but not

But

...and still

And

...or maybe

Or

...and both

Both

...neither one

Neither

...both ways

Ways

...one path

One

...many faces

Many

...single heart

Single

...multiplied by

Multiplied

...divided into

Divided

...made whole

Whole

...broken open

Open

...closed circle

Closed

...opens wide

Wide

...narrow focus

Narrow

...expands beyond

Beyond

...within reach

Within

...outside bounds

Outside

...inside out

Inside

...outside in

In

...out beyond

Out

...in deeper

Deeper

...surface calm

Surface

...depths below

Below

...above ground

Above

...below sky

Ground

...sky meets

Meets

...parts ways

Parts

...whole story

Story

...silence tells

Tells

...without words

Without

...with everything

With

...without need

Need

...want nothing

Want

...need everything

Everything

...nothing more

Nothing

...everything less

Less

...more enough

More

...less plenty

Plenty

...enough never

Never

...always when

When

...never now

Now

...when always

Always

...never then

Then

...always now...

Minimal Poetry Collection

Silence Speaks

...words fade...

The Last Drop

tears become...

Genesis

...light whispers...

Void

empty...full

Departure

...goodbye lingers

Dawn

darkness...surrender

Memory

...echoes remain...

Infinity

always...never...

Home

...heart returns

Truth

lies...then...

Title: ***Eternal Silence***
...whispers remain unheard...

Here are **100 profound micro-poems** (3 words or less) with ellipses and titles. Each evokes emotion, thought, or mystery:

---

1. **Echoes**
   ...voices drift away...
2. **Stillness**
   ...after the storm...
3. **Grief**
   ...never truly gone...
4. **Hope**
   ...light finds cracks...
5. **Cycle**
   ...again we begin...
6. **Alone**
   ...no one comes...
7. **Truth**
   ...beneath the silence...
8. **Ashes**
   ...something once burned...
9. **Dream**
   ...float beyond stars...
10. **Shadows**
    ...follow without sound...
11. **Time**
    ...never looks back...
12. **Rebirth**
    ...from the void...
13. **Linger**
    ...memories stay behind...
14. **Distance**
    ...hearts stretch thin...

15. **War**
...peace postponed again...
16. **Fading**
...we become echoes...
17. **Desire**
...never satisfied ever...
18. **Endings**
...always feel personal...
19. **Stars**
...watch us fall...
20. **Faith**
...without seeing, believing...
21. **Begin**
...again from nothing...
22. **Frozen**
...in a moment...
23. **Flesh**
...soft, yet temporary...
24. **Wounds**
...healed by time...
25. **Yearn**
...for what's lost...
26. **Collapse**
...under beauty's weight...
27. **Rise**
...after every fall...
28. **Unspoken**
...between our eyes...
29. **Guilt**
...hides in silence...
30. **Connection**
...when hands meet...

31. **Promise**
...words we trusted...
32. **Fade**
...into the blue...
33. **Bloom**
...despite the darkness...
34. **Pulse**
...rhythm of life...
35. **Tides**
...always returning home...
36. **Breathe**
...until breath ends...
37. **Ache**
...where love lived...
38. **Empty**
...without your voice...
39. **Light**
...after long dark...
40. **Drift**
...into forgetting...
41. **Fall**
...like golden leaves...
42. **Found**
...when least sought...
43. **Doubt**
...clouds the mind...
44. **Crack**
...beneath the surface...
45. **Sink**
...slow, then sudden...
46. **Walls**
...built from fear...

47. **Beg**
...for another chance...
48. **Pause**
...between heartbeats...
49. **Night**
...hides soft truths...
50. **Awake**
...but still dreaming...
51. **Roots**
...grow through pain...
52. **Reach**
...past what's gone...
53. **Burn**
...what we wrote...
54. **Rain**
...mends the earth...
55. **Silent**
...but not still...
56. **Hunger**
...never fed fully...
57. **Gone**
...before goodbye came...
58. **Thirst**
...for understanding...
59. **Leave**
...with no trace...
60. **Forgive**
...though not forgotten...
61. **Sing**
...even in sorrow...
62. **Cling**
...to faded warmth...

63. **Mist**
...shapes of memory...
64. **Bend**
...do not break...
65. **Shiver**
...under starlight's gaze...
66. **Mend**
...the broken parts...
67. **Inside**
...it still hurts...
68. **Speak**
...your truth softly...
69. **Let Go**
...of holding on...
70. **Return**
...to where began...
71. **Float**
...beyond the edge...
72. **Wash**
...the pain clean...
73. **Trust**
...despite the risk...
74. **Erase**
...but never fully...
75. **Break**
...to rebuild better...
76. **Move**
...with no map...
77. **Fallout**
...after the words...
78. **Glow**
...in ruined places...

79. **Alight**
...on trembling hope...
80. **Hide**
...from your reflection...
81. **Mirror**
...shatters the past...
82. **Wait**
...without knowing why...
83. **Crave**
...what wounds you...
84. **Forged**
...by silent fires...
85. **Knotted**
...in old memories...
86. **Choke**
...on unsaid things...
87. **Bleed**
...through the ink...
88. **Return**
...to dust again...
89. **Scar**
...proof of survival...
90. **Sink**
...into yourself...
91. **Gift**
...of being seen...
92. **Echo**
...your name still...
93. **Lurk**
...what you fear...
94. **Weep**
...without sound...

95. **Lie**
...with soft truth...
96. **Steal**
...a final glance...
97. **Trace**
...your fading face...
98. **Call**
...but no answer...
99. **Carve**
...hope in stone...
100. **Rest**
...in quiet light...

1. **Coffee**
...lifeblood begins here...
2. **Alarm**
...why so loud?
3. **Diet**
...starts tomorrow always
4. **Wi-Fi**
...please don't crash...
5. **Nap**
...just five minutes...
6. **TGIF**
...we survived!
7. **Pizza**
...never lets down
8. **Procrastinate**
...later sounds good

9. **Cat**
...owns the house
10. **Cat**
...still demands snacks
11. **Dog**
...breaks every rule
12. **Zoom**
...mute yourself please
13. **Meeting**
...could've been email
14. **Vacation**
...out of office
15. **Laundry**
...never ends though
16. **Email**
...spam gets smarter
17. **Boss**
...asks for “quick”
18. **Weekend**
...too short always
19. **Monday**
...ruins all vibes
20. **Plants**
...I kill gently
21. **GPS**
...lies every time
22. **Taco**
...holiday in shell
23. **Exercise**
...pain with purpose
24. **Gym**
...attendance optional, right?

25. **Scale**
...always betrays me
26. **Chocolate**
...my mood fixer
27. **Sponsor**
...needs more coffee
28. **To-do**
...list laughs silently
29. **Netflix**
...just one more
30. **Password**
...forgot it again
31. **Password**
..."P@ssw0rd123"? Really?
32. **Refund**
...never heard back
33. **Traffic**
...moves without me
34. **Queue**
...everyone's in front
35. **Hiccups**
...no sense at all
36. **Birthday**
...another year older
37. **UFO**
...still waiting proof
38. **Diet**
...the forbidden fruit
39. **Chocolate**
...dessert's best friend
40. **Pizza**
...makes math disappear

41. **Tax**
...unpaid anxiety
42. **Weather**
...changed its mind
43. **Streaming**
...buffering at climax
44. **Sleep**
...requires fewer thoughts
45. **Mondays**
...committee of doom
46. **Sunburn**
...pain with tan
47. **Vaccination**
...ouch, but safe
48. **Coffee**
...I'm not human
49. **Socks**
...missing their pair
50. **Pierogi**
...life's chewy gift
51. **Donut**
...hole in joy
52. **Cookie**
...crumbs are evidence
53. **Avocado**
...too expensive grief
54. **Onion**
...makes you cry
55. **Garlic**
...breath's best friend
56. **Caffeine**
...daily magic potion

57. **Password**
   ...must have symbols
58. **Drama**
   ...spreads fast though
59. **Smartphone**
   ...best distraction device
60. **Notifications**
   ...stress delivered instantly
61. **Ghosted**
   ...phone silent forever
62. **Dentist**
   ...pain in truth
63. **Budget**
   ...what's that again?
64. **Airport**
   ...carry-on only dreams
65. **Passport**
   ...lost at home
66. **Ticket**
   ...for air guitar
67. **Metabolism**
   ...takes vacation too
68. **Hair**
   ...decides wrong things
69. **Mirror**
   ...liar most days
70. **Gift**
   ...thanks, socks again
71. **Unicorn**
   ...stole my cake
72. **Dinosaur**
   ...roars silently today

73. **Robot**
...stole all jobs
74. **AI**
...tries small talk
75. **Selfie**
...deletes all filters
76. **Filter**
...makes me majestic
77. **Fashion**
...follows no logic
78. **Smile**
...dentist approved sign
79. **Kid**
...makes silence loud
80. **Toddler**
...tiny boss tantrum
81. **Teenager**
...answers with grunts
82. **Parenting**
...coffee earns degrees
83. **Zombie**
...still didn't vote
84. **Alien**
...offers bad Wi-Fi
85. **Weekend**
...disappears too fast
86. **Holiday**
...worth the chaos
87. **Budget**
...spent yesterday somewhere
88. **Coupon**
...never saves enough

89. **Deal**
...ends at checkout
90. **Refund**
...denied politely sometimes
91. **Delivery**
...ran into building
92. **Takeout**
...smarter than cooking
93. **Meal**
...ended in snacks
94. **Recipe**
...ignored last sentence
95. **Ingredient**
...missing again always
96. **Spice**
...set mouth ablaze
97. **Ice-cream**
...melts urgent gifts
98. **Brainfreeze**
...worth every scoop
99. **Chocolate**
...makes math melt
100. **Diet**
...surrenders today again
101. **Procrastinate**
...master of delay
102. **Champion**
...of doing nothing
103. **Effort**
...pauses indefinitely often
104. **Focus**
...sprints then quits

105. **Concentration**
...takes coffee breaks
106. **Miracle**
...found missing socks
107. **Legend**
...in my mirror
108. **Cereal**
...counts as breakfast
109. **Cereal**
...counts as dinner
110. **Cereal**
...counts as life
111. **Noodles**
...instant happiness now
112. **Ice**
...melted my resolve
113. **Heat**
...makes brain liquefy
114. **Sweat**
...instantly regrets shirt
115. **Gym**
...I ghost often
116. **Weights**
...still lifting lists
117. **Spin**
...on leader of stairs
118. **Yoga**
...makes me wobble
119. **Zen**
...in chaos seldom
120. **Peace**
...interrupts with thoughts

121. **Meditate**
...mind stays busy
122. **Relax**
...later, not now
123. **Vacation**
...requires packing stress
124. **Suitcase**
...overfilled again, sigh
125. **Airport**
...security judged socks
126. **Ticket**
...lost boarding time
127. **Luggage**
...thinks it's free
128. **Hotel**
...keys never work
129. **Pillow**
...still flat morning
130. **Sleep**
...avoids my attempts
131. **Dream**
...invent bizarre plots
132. **Alarm**
...betrays restful nights
133. **Shower**
...makes me think
134. **Shampoo**
...lots of bubbles
135. **Mirror**
...judges bad hair
136. **Outfit**
...made wrong choices

137. **Date**

...texted “seen”?

138. **Swipe**

...left again instantly

139. **Match**

...still unmatched me

140. **Crush**

...says hi never

141. **Coffee**

...need more now

142. **Tea**

...too polite hot

143. **Juice**

...thirst pretender only

144. **Water**

...I adore you

145. **Calories**

...uninvited guests daily

146. **Snack**

...secret midnight agent

147. **Cookie**

...crumbs as proof

148. **Cheese**

...makes life cheesier

149. **Pasta**

...swallows all forks

150. **Naan**

...bread with attitude

151. **Soup**

...liquid hug always

152. **Stew**

...future leftovers today

153. **Curry**
...spice tests courage
154. **Steak**
...charred love story
155. **Burger**
...greasy happy heart
156. **Salad**
...crunch feels healthy
157. **Kale**
...needs redemption soon
158. **Avocado**
...brings toast alive
159. **Toast**
...blackened every time
160. **Eggs**
...breakfast's best friends
161. **Oatmeal**
...gives fuzzy comfort
162. **Yogurt**
...tangy life boost
163. **Smoothie**
...masked vegetable trick
164. **Milk**
...cow's gift daily
165. **Cheese**
...makes crackers dance
166. **Bread**
...life's golden circle
167. **Bagel**
...hole yields meaning
168. **Donut**
...round joy delivered

169. **Cake**
...birthday's best excuse
170. **Pie**
...fruit's baked hug
171. **Candy**
...little sugar bombs
172. **Chocolate**
...my evil twin
173. **Ice-cream**
...cold joy explosion
174. **Cookie**
...sweet guilt trip
175. **Brownie**
...fudge party inside
176. **Candy**
...makes dentist cry
177. **Lollipop**
...infinite sugar pain
178. **Gum**
...sticky regrets later
179. **Bubblegum**
...balloon mouth hazard
180. **Mint**
...fresh breath saver
181. **Chewing**
...class distraction tool
182. **Popcorn**
...movie's sidekick snack
183. **Chips**
...crunch therapy sessions
184. **Pretzel**
...twisted snack art

185. **Cracker**
...cracker comically bland
186. **Nacho**
...cheese disaster waiting
187. **Salsa**
...spicy chaos contained
188. **Guac**
...never enough chips
189. **Tortilla**
...wraps all issues
190. **Burrito**
...rolls happiness inside
191. **Taco**
...joy in shell
192. **Enchilada**
...saucy mood bomb
193. **Quesadilla**
...cheesy fold heaven
194. **Salad**
...just leaves mostly
195. **Soup**
...liquid life balm
196. **Burger**
...hands get greasy
197. **Fries**
...dip me deeper
198. **Onion**
...rain-making vegetable
199. **Garlic**
...breath persists forever
200. **Pepper**
...tiny spice punch

Hope these bring a **smile and a chuckle**! Want themed edits—romance, sci-fi, cynical? Just let me know!

Echoes of Silence ...forgotten breath...

Beneath the Void ...nothing remains...

Tides of Memory drift away...

Ashes Speak ...burned truth...

Whispered Time ...never now...

Cosmic Thread ...woven stars...

Fractured Reflection ...who was I...

Still Becoming ...rooted change...

Crimson Truth ...bleeds quietly...

Ghosts of Then ...once again...

Fading Light ...almost gone...

Thought Fugue ...forgot to think...

Hollow Pulse ...love left...

Dream Residue ...never real...

🧭 **Lost North** ...no direction...

💭 Existence & Identity

1. **Self Echo** ...who am I...
2. **Unwritten Name** ...no label...
3. **Becoming** ...still forming...
4. **Skin Memory** ...felt before...
5. **Silent Mirror** ...no reflection...
6. **Inevitable Shift** ...change me...
7. **Unclaimed Soul** ...not mine...
8. **Core Silence** ...deep hush...
9. **Hidden Pulse** ...beneath skin...
10. **Flicker Self** ...almost real...

🌌 Cosmos & Time

11. **Star Whisper** ...light speaks...
12. **Time Fold** ...again now...
13. **Gravity Dream** ...pull me...
14. **Orbit Lost** ...no center...
15. **Moon Thought** ...silver mind...
16. **Eternal Drift** ...never ends...
17. **Black Hole** ...swallowed whole...
18. **Solar Ache** ...burn slow...
19. **Timeless Dust** ...always was...
20. **Cosmic Breath** ...space sighs...

🌿 Nature & Earth

21. **Rooted Silence** ...grow still...

22. **Rain Memory** ...fall again...
23. **Wind Truth** ...whispers loud...
24. **Stone Sleep** ...rest deep...
25. **Leaf Pulse** ...green beat...
26. **River Thought** ...flow on...
27. **Forest Echo** ...lost sound...
28. **Sun Fade** ...light gone...
29. **Mountain Wait** ...stand still...
30. **Ocean Speak** ...blue voice...

🩶 Love & Longing

31. **Heart Ghost** ...felt once...
32. **Unspoken Want** ...say it...
33. **Touchless Love** ...reach far...
34. **Crushed Bloom** ...petals fall...
35. **Warm Absence** ...still here...
36. **Soul Thread** ...tied tight...
37. **Kiss Echo** ...linger long...
38. **Fever Dream** ...burn slow...
39. **Empty Arms** ...held nothing...
40. **Love Residue** ...never left...

Loss & Grief

41. **Ash Whisper** ...gone now...
42. **Final Breath** ...last sigh...
43. **Grief Song** ...silent scream...
44. **Buried Light** ...dimmed forever...
45. **Gone Again** ...still missing...
46. **Tear Memory** ...salt remains...
47. **Shadow Hug** ...dark comfort...

48. **Echo Tomb** ...heard once...
49. **Fading Name** ...forgotten sound...
50. **Cold Goodbye** ...no warmth...

🧠 Thought & Mind

51. **Idea Ghost** ...almost formed...
52. **Mind Fog** ...blurred thought...
53. **Dream Loop** ...again again...
54. **Mental Drift** ...float far...
55. **Thought Silence** ...no words...
56. **Brain Pulse** ...tick tick...
57. **Inner Storm** ...quiet chaos...
58. **Memory Flicker** ...half remembered...
59. **Logic Break** ...fractured sense...
60. **Mental Thread** ...unraveling now...

🧘 Spiritual & Abstract

61. **Soul Silence** ...listen deep...
62. **Prayer Dust** ...whispered hope...
63. **Sacred Void** ...holy absence...
64. **Divine Ache** ...want more...
65. **Faith Flicker** ...almost belief...
66. **Spirit Drift** ...float free...
67. **Sacrifice Echo** ...gave all...
68. **Sacred Breath** ...inhale light...
69. **Holy Wait** ...stillness speaks...
70. **Blessed Fade** ...grace gone...

🔥 Emotion & Intensity

71. **Rage Whisper** ...burn quiet...
72. **Joy Shiver** ...felt fast...
73. **Fear Pulse** ...thump thump...
74. **Hope Thread** ...hold tight...
75. **Doubt Bloom** ...grows slow...
76. **Anger Fog** ...red blur...
77. **Peace Ache** ...want still...
78. **Desire Flame** ...never out...
79. **Shame Dust** ...settled deep...
80. **Bliss Fade** ...almost joy...

🧳 Journey & Change

81. **Path Lost** ...no map...
82. **Crossroads Wait** ...choose now...
83. **Wander Pulse** ...move again...
84. **Exit Sign** ...leave now...
85. **Arrival Ghost** ...never came...
86. **Step Echo** ...walked before...
87. **Direction Blur** ...where now...
88. **Journey Thread** ...follow me...
89. **Departure Light** ...gone soon...
90. **Change Whisper** ...shift slow...

🕶 Mystery & Surreal

91. **Dream Static** ...flicker real...
92. **Fog Thought** ...unclear truth...
93. **Shadow Speak** ...dark voice...
94. **Illusion Thread** ...almost real...
95. **Mask Pulse** ...hide me...
96. **Secret Drift** ...float hidden...

97. **Hidden Flame** ...burn unseen...
98. **Unknown Ache** ...felt deep...
99. **Veil Silence** ...covered truth...
100. **Mystery Breath** ...inhale wonder...

These are designed to evoke emotion, spark introspection, and invite interpretation.

Existence & Identity

1. **Unlabeled Soul** ...just being...
2. **Inner Echo** ...me again...
3. **Identity Blur** ...who now...
4. **Silent Core** ...deep within...
5. **Becoming Still** ...almost me...
6. **Self Unfolding** ...open slowly...
7. **Name Forgotten** ...called once...
8. **Truth Hidden** ...beneath skin...
9. **I Am** ...nothing else...
10. **Mask Off** ...bare face...

Cosmos & Time

11. **Starless Sky** ...no light...
12. **Time Pause** ...hold still...
13. **Orbit Broken** ...lost path...
14. **Moonless Night** ...dark silence...
15. **Cosmic Wait** ...forever now...
16. **Galactic Drift** ...float far...
17. **Solar Silence** ...burn quiet...
18. **Timeless Pulse** ...always ticking...
19. **Space Ache** ...vast longing...
20. **Nebula Thought** ...clouded mind...

🌿 Nature & Earth

21. **Root Ache** ...grow slow...
22. **Rain Speak** ...soft truth...
23. **Wind Pause** ...held breath...
24. **Stone Memory** ...never moved...
25. **Leaf Drift** ...fall again...
26. **River Wait** ...flow still...
27. **Forest Sleep** ...green hush...
28. **Sunless Day** ...dim light...
29. **Mountain Thought** ...stand firm...
30. **Ocean Silence** ...deep sound...

🩶 Love & Longing

31. **Heart Wait** ...still yours...
32. **Unheld Touch** ...reach far...
33. **Love Ghost** ...felt once...
34. **Warm Fade** ...almost here...
35. **Soul Ache** ...missed you...
36. **Kiss Lost** ...gone now...
37. **Tender Drift** ...soft goodbye...
38. **Crushed Hope** ...bloom died...
39. **Empty Hug** ...held air...
40. **Love Thread** ...tied tight...

🕯️ Loss & Grief

41. **Ash Drift** ...gone again...
42. **Final Light** ...dimmed now...
43. **Grief Pulse** ...beat slow...

44. **Buried Name** ...forgotten sound...
45. **Gone Silence** ...no echo...
46. **Tear Dust** ...salt remains...
47. **Shadow Wait** ...dark comfort...
48. **Echo Fade** ...heard once...
49. **Cold Absence** ...no warmth...
50. **Lost Again** ...still missing...

🧠 Thought & Mind

51. **Idea Flicker** ...almost formed...
52. **Mind Blur** ...unclear now...
53. **Dream Pulse** ...tick slow...
54. **Mental Drift** ...float far...
55. **Thought Pause** ...no words...
56. **Brain Fog** ...blurred sense...
57. **Inner Storm** ...quiet chaos...
58. **Memory Thread** ...unraveling now...
59. **Logic Break** ...fractured truth...
60. **Mental Silence** ...still thinking...

🧘 Spiritual & Abstract

61. **Soul Wait** ...listen deep...
62. **Prayer Whisper** ...soft hope...
63. **Sacred Ache** ...holy longing...
64. **Faith Drift** ...almost belief...
65. **Spirit Pulse** ...float free...
66. **Sacrifice Breath** ...gave all...
67. **Divine Silence** ...grace gone...
68. **Blessed Wait** ...stillness speaks...
69. **Holy Thread** ...tied tight...

70. **Sacred Fade** ...light dimmed...

🔥 Emotion & Intensity

71. **Rage Pulse** ...burn slow...
72. **Joy Ache** ...felt fast...
73. **Fear Drift** ...float dark...
74. **Hope Thread** ...hold tight...
75. **Doubt Bloom** ...grows deep...
76. **Anger Fog** ...red blur...
77. **Peace Ache** ...want still...
78. **Desire Flame** ...never out...
79. **Shame Dust** ...settled deep...
80. **Bliss Fade** ...almost joy...

🧳 Journey & Change

81. **Path Drift** ...no map...
82. **Crossroads Ache** ...choose now...
83. **Wander Pulse** ...move again...
84. **Exit Thought** ...leave now...
85. **Arrival Ghost** ...never came...
86. **Step Echo** ...walked before...
87. **Direction Blur** ...where now...
88. **Journey Thread** ...follow me...
89. **Departure Light** ...gone soon...
90. **Change Whisper** ...shift slow...

🕶️ Mystery & Surreal

91. **Dream Static** ...flicker real...

92. **Fog Thought** ...unclear truth...
93. **Shadow Speak** ...dark voice...
94. **Illusion Thread** ...almost real...
95. **Mask Pulse** ...hide me...
96. **Secret Drift** ...float hidden...
97. **Hidden Flame** ...burn unseen...
98. **Unknown Ache** ...felt deep...
99. **Veil Silence** ...covered truth...
100. **Mystery Breath** ...inhale wonder...

🧠 Existential Absurdity

1. **Existence Shrugged** ...why bother...
2. **Deep-ish Thought** ...maybe not...
3. **Philosophy Lite** ...think less...
4. **Being Tired** ...again me...
5. **Soul Buffering** ...loading self...
6. **Identity Crisis** ...who dis...
7. **Unbecoming Me** ...oops again...
8. **Existential Nope** ...not today...
9. **Self-ish** ...just me...
10. **Conscious-ish** ...half awake...

☕ Daily Life Chaos

11. **Coffee Gospel** ...sip salvation...
12. **Alarm Betrayal** ...not yet...
13. **Laundry Saga** ...still dirty...
14. **To-Don't List** ...do nothing...
15. **Snack Philosophy** ...eat feelings...
16. **Inbox Horror** ...still full...
17. **Zoom Doom** ...mute me...

18. **Traffic Zen** ...rage calmly...
19. **Dish Drama** ...again sink...
20. **Sleep Tease** ...almost nap...

Animals Being Weird

21. **Cat Logic** ...sit there...
22. **Dog Thoughts** ...food now...
23. **Bird Gossip** ...tweet drama...
24. **Fish Exist** ...just swim...
25. **Cow Vibes** ...moo softly...
26. **Goat Chaos** ...climb everything...
27. **Horse Secrets** ...neigh quietly...
28. **Duck Agenda** ...quack plans...
29. **Lizard Chill** ...sun nap...
30. **Bee Business** ...buzz buzz...

Love & Dating Fails

31. **Crush Panic** ...don’t text...
32. **Flirt Malfunction** ...wink fail...
33. **Romantic-ish** ...kind of...
34. **Ghosted Again** ...who me...
35. **Love Math** ...carry one...
36. **Date Escape** ...check please...
37. **Heart Buffering** ...loading love...
38. **Cuddle Tax** ...pay snuggles...
39. **Breakup Snack** ...eat everything...
40. **Swipe Regret** ...left again...

Fake Deep Wisdom

41. **Zen-ish** ...almost peace...
42. **Chakra Confused** ...which one...
43. **Meditation Fail** ...fell asleep...
44. **Aura Drama** ...too bright...
45. **Crystal Doubt** ...maybe magic...
46. **Vibe Check** ...failed again...
47. **Energy Meh** ...low charge...
48. **Spiritual-ish** ...kind of...
49. **Mantra Panic** ...forgot words...
50. **Breathe-ish** ...half inhale...

Food & Drink Nonsense

51. **Snack Attack** ...eat now...
52. **Fridge Stare** ...still empty...
53. **Spicy Regret** ...mouth fire...
54. **Soup Existential** ...just broth...
55. **Pizza Gospel** ...cheese truth...
56. **Fork Betrayal** ...stabbed self...
57. **Muffin Mystery** ...where top...
58. **Toast Drama** ...burnt again...
59. **Soda Wisdom** ...pop fizz...
60. **Ice Cream Therapy** ...cry scoop...

Tech & Internet Woes

61. **WiFi Ghost** ...gone again...
62. **Battery Panic** ...low now...
63. **App Regret** ...downloaded shame...
64. **Autocorrect War** ...duck you...
65. **Scroll Spiral** ...never end...
66. **TikTok Timewarp** ...lost hours...

67. **Emoji Debate** ...which face...
68. **Cloud Confusion** ...where files...
69. **Bluetooth Betrayal** ...won’t connect...
70. **AI Side-Eye** ...judging me...

🛌 Sleep & Laziness

71. **Nap Philosophy** ...sleep now...
72. **Bed Loyalty** ...never leave...
73. **Dream Chaos** ...what was...
74. **Pillow Politics** ...fluff war...
75. **Blanket Trap** ...stuck again...
76. **Snooze Gospel** ...ten more...
77. **Sleep Math** ...never enough...
78. **Rest-ish** ...almost nap...
79. **Yawn Ritual** ...repeat often...
80. **Sleep Olympics** ...gold nap...

🎭 Social Awkwardness

81. **Introvert Anthem** ...stay home...
82. **Small Talk** ...weather again...
83. **Eye Contact** ...look away...
84. **Party Panic** ...hide now...
85. **Handshake Fail** ...missed again...
86. **Awkward Silence** ...say something...
87. **Laugh Buffer** ...too late...
88. **Exit Strategy** ...leave early...
89. **Group Text** ...who’s that...
90. **Social-ish** ...kind of...

🧹 Household Chaos

91. **Dust Rebellion** ...never gone...
92. **Vacuum Rage** ...scream floor...
93. **Sock Mystery** ...one gone...
94. **Closet War** ...hang chaos...
95. **Trash Drama** ...overflow again...
96. **Light Switch** ...wrong one...
97. **Door Creak** ...ghost maybe...
98. **Laundry Lies** ...still damp...
99. **Pan Grease** ...won't leave...
100. **Remote Hide** ...lost again...

🎨 Art & Creativity

101. **Paint Panic** ...wrong color...
102. **Poem-ish** ...kind of...
103. **Brush Drama** ...too wet...
104. **Canvas Regret** ...start over...
105. **Sculpt Fail** ...blob now...
106. **Creative Fog** ...no idea...
107. **Muse Ghost** ...gone again...
108. **Color War** ...clash hard...
109. **Art-ish** ...almost good...
110. **Doodle Gospel** ...line truth...

🎮 Gaming & Nerd Life

111. **Level Rage** ...died again...
112. **Loot Regret** ...wrong chest...
113. **Boss Panic** ...run now...

114. **Quest Confused** ...where map...
115. **XP Ache** ...grind forever...
116. **Controller Betrayal** ...wrong button...
117. **Lag Doom** ...freeze now...
118. **Save Fail** ...lost progress...
119. **Game Over** ...again again...
120. **Victory Dance** ...awkward wiggle...

Weird & Whimsical

121. **Vampire Nap** ...sun bad...
122. **Ghost Text** ...who boo...
123. **Zombie Thoughts** ...need brain...
124. **Alien Vibes** ...beam me...
125. **Witch Snack** ...hex chips...
126. **Dragon Yawn** ...fire breath...
127. **Mermaid Drama** ...wet again...
128. **Fairy Taxes** ...pay glitter...
129. **Goblin Mood** ...steal socks...
130. **Wizard WiFi** ...magic signal...

Random Absurdity

131. **Sock Philosophy** ...one gone...
132. **Chair Loyalty** ...sit forever...
133. **Ceiling Thoughts** ...look up...
134. **Toaster Rage** ...burnt again...
135. **Window Gossip** ...see everything...
136. **Lamp Mood** ...dim vibe...
137. **Pen Betrayal** ...won't write...
138. **Mug Identity** ...not yours...
139. **Stair Regret** ...too many...

140. **Fan Drama** ...blow weird...

More Food Nonsense

141. **Banana Betrayal** ...bruised again...
142. **Soup Existential** ...just broth...
143. **Cheese Gospel** ...holy snack...
144. **Pickle Panic** ...too sour...
145. **Bread Mood** ...crusty vibe...
146. **Egg Mystery** ...which came...
147. **Spoon Drama** ...bent again...
148. **Cake Whisper** ...eat now...
149. **Taco Truth** ...folded wisdom...
150. **Cereal Chaos** ...milk first...

Celebration & Chaos

151. **Birthday Panic** ...forgot again...
152. **Party Fog** ...too loud...
153. **Gift Regret** ...wrong one...
154. **
155. deeply profound poetry with a title and 3 or less words with ellipsis
156. **Whispers...**
157. A fleeting breath, a silent plea, The universe in you, in me. Echoes fade, but truth remains, Love's ancient song through sun and rains.
158. 3 or less words making profound poetry add ellipsis to end or beginnning and add a title
159. **Cosmic Dust...**
160. Silent. Vast. Forever...

Soul...

Deep. Still. Boundless...

Echoes...

Past. Present. Future...

Becoming...

Bloom. Fade. Return...

Eternity...

Now. Always. Never...

Truth...

See. Feel. Know...

Silence...

Hush. Breathe. Be...

Light...

Dawn. Day. Dusk...

Wisdom...

Learn. Grow. Share...

Wonder...

Gaze. Awe. Dream...

Freedom...

Fly. Soar. Unbound...

191. ______________________________

192. Hope...

193. Seed. Sprout. Rise...

194. ______________________________

195. Fate...

196. Spin. Weave. End...

197. ______________________________

198. Peace...

199. Calm. Rest. Serene...

200. ______________________________

201. Change...

202. Flow. Shift. Transform...

203. ______________________________

204. Mystery...

205. Seek. Find. Lost...

206. ______________________________

207. Journey...

208. Path. Step. Ahead...

209. ______________________________

210. Dream...

211. Sleep. Wake. Imagine...

212. ______________________________

213. Home...

214. Heart. Soul. Place...

215. ______________________________

216. Memory...

217. Recall. Glimpse. Hold...

218. ______________________________

219. Loss...

220. Grief. Empty. Echo...

221. ______________________________

222. Strength...

223. Rise. Stand. Endure...

224. ______________________________

225. Grace...

226. Gift. Flow. Receive...

227. ______________________________

228. Voice...

229. Speak. Sing. Hear...

230.

231. Spirit...

232. Essence. Spark. Alive...

233.

234. Embrace...

235. Hold. Feel. Connect...

236.

237. Unseen...

238. Sense. Trust. Beyond...

239.

240. Moment...

241. Here. Now. Gone...

242.

243. Void...

244. Dark. Deep. Space...

245.

246. Origin...

247. First. Start. Birth...

248.

249. Ending...

250. Close. Cease. Finish...

Art...

Create. Express. Feel...

Self...

Find. Know. Be...

Becoming...

Grow. Evolve. Soar...

Illusion...

See. Believe. Dispel...

Rhythm...

Beat. Pulse. Life...

Union...

Join. Merge. One...

Bloom...

Open. Unfold. Radiate...

Quest...

Seek. Search. Discover...

Flow...

Move. Drift. Guide...

Anchor...

Hold. Ground. Secure...

Glimmer...
Hope. Light. Persist...

Anchor...
Root. Hold. Steadfast...

Unfurl...
Open. Bloom. Expand...

Whisper...
Soft. Secret. Hear...

Echoes...
Past. Present. Future...

Becoming...
Grow. Evolve. Soar...

Eternity...
Now. Always. Never...

Truth...
See. Feel. Know...

Silence...
Hush. Breathe. Be...

Light...
Dawn. Day. Dusk...

311. ____________________

312. Wisdom...

313. Learn. Grow. Share...

314. ____________________

315. Wonder...

316. Gaze. Awe. Dream...

317. ____________________

318. Freedom...

319. Fly. Soar. Unbound...

320. ____________________

321. Hope...

322. Seed. Sprout. Rise...

323. ____________________

324. Fate...

325. Spin. Weave. End...

326. ____________________

327. Peace...

328. Calm. Rest. Serene...

329. ____________________

330. Change...

331. Flow. Shift. Transform...

332. ____________________

333. Mystery...

334. Seek. Find. Lost...

335. ____________________

336. Journey...

337. Path. Step. Ahead...

338. ____________________

339. Dream...

340. Sleep. Wake. Imagine...

341. ______________________________

342. Home...

343. Heart. Soul. Place...

344. ______________________________

345. Memory...

346. Recall. Glimpse. Hold...

347. ______________________________

348. Loss...

349. Grief. Empty. Echo...

350. ______________________________

351. Strength...

352. Rise. Stand. Endure...

353. ______________________________

354. Grace...

355. Gift. Flow. Receive...

356. ______________________________

357. Voice...

358. Speak. Sing. Hear...

359. ______________________________

360. Spirit...

361. Essence. Spark. Alive...

362. ______________________________

363. Embrace...

364. Hold. Feel. Connect...

365. ______________________________

366. Unseen...

367. Sense. Trust. Beyond...

368. ______________________________

369. Moment...

370. Here. Now. Gone...

371. ______________________________

372. Void...

373. Dark. Deep. Space...

374. ______________________________

375. Origin...

376. First. Start. Birth...

377. ______________________________

378. Ending...

379. Close. Cease. Finish...

380. ______________________________

381. Art...

382. Create. Express. Feel...

383. ______________________________

384. Self...

385. Find. Know. Be...

386. ______________________________

387. Bloom...

388. Open. Unfold. Radiate...

389. ______________________________

390. Quest...

391. Seek. Search. Discover...

392. ______________________________

393. Flow...

394. Move. Drift. Guide...

395. ______________________________

396. Anchor...

397. Hold. Ground. Secure...

398. ______________________________

399. Pause...

400. Breathe. Wait. Reflect...

401. ____________________

402. Spark...

403. Ignite. Shine. Inspire...

404. ____________________

405. Dance...

406. Move. Rhythm. Joy...

407. ____________________

408. Bloom...

409. Open. Unfold. Radiate...

410. ____________________

411. Fade...

412. Dim. Soft. Gone...

413. ____________________

414. Weave...

415. Spin. Thread. Connect...

416. ____________________

417. Portal...

418. Open. Pass. Beyond...

419. ____________________

420. Bloom...

421. Open. Unfold. Radiate...

422. ____________________

423. Pulse...

424. Beat. Live. Throb...

425. ____________________

426. Veil...

427. Hide. Reveal. Peer...

428. ____________________

429. Shift...

430. Move. Change. Become...

431. ______________________________
432. Ascent...
433. Climb. Rise. Soar...

434. ______________________________
435. Descent...
436. Fall. Sink. Deep...

437. ______________________________
438. Mirror...
439. Reflect. See. Soul...

440. ______________________________
441. Thread...
442. Fine. Bind. Connect...

443. ______________________________
444. Abyss...
445. Deep. Dark. Fall...

446. ______________________________
447. Canvas...
448. Blank. Paint. Create...

449. ______________________________
450. Root...
451. Ground. Deep. Grow...

452. ______________________________
453. Edge...
454. Brink. Limit. Dare...

455. ______________________________
456. Unseen...
457. Sense. Feel. Beyond...

458. ______________________________
459. Echo...
460. Repeat. Reverb. Fade...

461. ____________________

462. Shift...

463. Move. Change. Flow...

464. ____________________

465. Quest...

466. Seek. Search. Find...

467. ____________________

468. Gate...

469. Open. Close. Pass...

470. ____________________

471. Core...

472. Heart. Center. Essence...

473. ____________________

474. Tide...

475. Rise. Fall. Flow...

476. ____________________

477. Dream...

478. Sleep. Wake. Imagine...

479. ____________________

480. Glimpse...

481. See. Brief. Flash...

482. ____________________

483. Unfold...

484. Open. Reveal. Grow...

485. ____________________

486. Silence...

487. Still. Calm. Peace...

488. ____________________

489. Horizon...

490. Meet. Far. Beyond...

491. ______________________________

492. Portal...

493. Open. Pass. New...

494. ______________________________

495. Whisper...

496. Soft. Hint. Know...

497. ______________________________

498. Canvas...

499. Blank. Paint. Life...

500. ______________________________

501. Vessel...

502. Hold. Carry. Soul...

503. ______________________________

504. Drift...

505. Float. Wander. Calm...

506. ______________________________

507. Spark...

508. Tiny. Bright. Life...

509. ______________________________

510. Echo...

511. Repeat. Resound. Time...

512. ______________________________

513. Shift...

514. Move. Change. Evolve...

515. ______________________________

516. Web...

517. Intertwine. Connect. Bind...

518. ______________________________

519. Root...

520. Deep. Ground. Begin...

521. ______________________________

522. Bloom...

523. Open. Unfold. Radiate...

524. ______________________________

525. Veil...

526. Hide. Reveal. Truth...

527. ______________________________

528. Abyss...

529. Vast. Empty. Deep...

530. ______________________________

531. Pulse...

532. Beat. Life. Rhythm...

533. ______________________________

534. Unfurl...

535. Open. Release. Fly...

536. ______________________________

537. Core...

538. Center. Heart. Being...

539. ______________________________

540. Tide...

541. Push. Pull. Cycle...

542. ______________________________

543. Gleam...

544. Shine. Catch. Light...

545. ______________________________

546. Portal...

547. Enter. Exit. Realm...

548. ______________________________

549. Whisper...

550. Soft. Gentle. Voice...

Canvas...

Create. Paint. Dream...

Vessel...

Hold. Carry. Journey...

Drift...

Float. Flow. Release...

Spark...

Ignite. Kindle. Grow...

Echo...

Faint. Distant. Call...

Shift...

Turn. Alter. Path...

Web...

Catch. Entwine. Form...

Root...

Anchor. Connect. Life...

Whispers...

A fleeting breath, a silent plea, The universe in you, in me. Echoes fade, but truth remains, Love's ancient song through sun and rains.

Cosmic Dust...

Silent. Vast. Forever...

Soul...

Deep. Still. Boundless...

Echoes...

Past. Present. Future...

Becoming...

Bloom. Fade. Return...

Eternity...

Now. Always. Never...

Truth...

See. Feel. Know...

Silence...

Hush. Breathe. Be...

Light...

Dawn. Day. Dusk...

Wisdom...

Learn. Grow. Share...

Wonder...

Gaze. Awe. Dream...

Freedom...

Fly. Soar. Unbound...

Hope...

Seed. Sprout. Rise...

Fate...

Spin. Weave. End...

Peace...

Calm. Rest. Serene...

Change...

Flow. Shift. Transform...

Mystery...

Seek. Find. Lost...

Journey...

Path. Step. Ahead...

Dream...

Sleep. Wake. Imagine...

Home...

Heart. Soul. Place...

Memory...

Recall. Glimpse. Hold...

Loss...

Grief. Empty. Echo...

Strength...

Rise. Stand. Endure...

Grace...

Gift. Flow. Receive...

Voice...

Speak. Sing. Hear...

Spirit...

Essence. Spark. Alive...

Embrace...

Hold. Feel. Connect...

Unseen...

Sense. Trust. Beyond...

Moment...

Here. Now. Gone...

Void...

Dark. Deep. Space...

Origin...

First. Start. Birth...

Ending...

Close. Cease. Finish...

Art...

Create. Express. Feel...

Self...

Find. Know. Be...

Becoming...

Grow. Evolve. Soar...

Illusion...

See. Believe. Dispel...

Rhythm...

Beat. Pulse. Life...

Union...

Join. Merge. One...

Bloom...

Open. Unfold. Radiate...

Quest...

Seek. Search. Discover...

Flow...

Move. Drift. Guide...

Anchor...

Hold. Ground. Secure...

Glimmer...

Hope. Light. Persist...

Anchor...

Root. Hold. Steadfast...

Unfurl...

Open. Bloom. Expand...

Whisper...

Soft. Secret. Hear...

Echoes...

Past. Present. Future...

Becoming...

Grow. Evolve. Soar...

Eternity...

Now. Always. Never...

Truth...

See. Feel. Know...

Silence...

Hush. Breathe. Be...

Light...

Dawn. Day. Dusk...

Wisdom...

Learn. Grow. Share...

Wonder...

Gaze. Awe. Dream...

Freedom...

Fly. Soar. Unbound...

Hope...

Seed. Sprout. Rise...

Fate...

Spin. Weave. End...

Peace...

Calm. Rest. Serene...

Change...

Flow. Shift. Transform...

Mystery...

Seek. Find. Lost...

Journey...

Path. Step. Ahead...

Dream...

Sleep. Wake. Imagine...

Home...

Heart. Soul. Place...

Memory...

Recall. Glimpse. Hold...

Loss...

Grief. Empty. Echo...

Strength...

Rise. Stand. Endure...

Grace...

Gift. Flow. Receive...

Voice...

Speak. Sing. Hear...

Spirit...

Essence. Spark. Alive...

Embrace...

Hold. Feel. Connect...

Unseen...

Sense. Trust. Beyond...

Moment...

Here. Now. Gone...

Void...

Dark. Deep. Space...

Origin...

First. Start. Birth...

Ending...

Close. Cease. Finish...

Art...

Create. Express. Feel...

Self...

Find. Know. Be...

Bloom...

Open. Unfold. Radiate...

Quest...

Seek. Search. Discover...

Flow...

Move. Drift. Guide...

Anchor...

Hold. Ground. Secure...

Pause...

Breathe. Wait. Reflect...

Spark...

Ignite. Shine. Inspire...

Dance...

Move. Rhythm. Joy...

Bloom...

Open. Unfold. Radiate...

Fade...

Dim. Soft. Gone...

Weave...

Spin. Thread. Connect...

Portal...

Open. Pass. Beyond...

Bloom...

Open. Unfold. Radiate...

Pulse...

Beat. Live. Throb...

Veil...

Hide. Reveal. Peer...

Shift...

Move. Change. Become...

Ascent...

Climb. Rise. Soar...

Descent...

Fall. Sink. Deep...

Mirror...

Reflect. See. Soul...

Thread...

Fine. Bind. Connect...

Abyss...

Deep. Dark. Fall...

Canvas...

Blank. Paint. Create...

Root...

Ground. Deep. Grow...

Edge...

Brink. Limit. Dare...

Unseen...

Sense. Feel. Beyond...

Echo...

Repeat. Reverb. Fade...

Shift...

Move. Change. Flow...

Quest...

Seek. Search. Find...

3 or less words making profound poetry add ellipsis to end or beginnning and add a title

---

Gate...

Open. Close. Pass...

Core...

Heart. Center. Essence...

Tide...

Rise. Fall. Flow...

Dream...

Sleep. Wake. Imagine...

Glimpse...

See. Brief. Flash...

Unfold...

Open. Reveal. Grow...

Silence...

Still. Calm. Peace...

Horizon...

Meet. Far. Beyond...

Portal...

Open. Pass. New...

Whisper...

Soft. Hint. Know...

Canvas...

Blank. Paint. Life...

Vessel...

Hold. Carry. Soul...

Drift...

Float. Wander. Calm...

Spark...

Tiny. Bright. Life...

Echo...

Repeat. Resound. Time...

Shift...

Move. Change. Evolve...

Web...

Intertwine. Connect. Bind...

Root...

Deep. Ground. Begin...

Bloom...

Open. Unfold. Radiate...

Veil...

Hide. Reveal. Truth...

Abyss...

Vast. Empty. Deep...

Pulse...

Beat. Life. Rhythm...

Unfurl...

Open. Release. Fly...

Core...

Center. Heart. Being...

Tide...

Push. Pull. Cycle...

Gleam...

Shine. Catch. Light...

Portal...

Enter. Exit. Realm...

Whisper...

Soft. Gentle. Voice...

Canvas...

Create. Paint. Dream...

Vessel...

Hold. Carry. Journey...

Drift...

Float. Flow. Release...

Spark...

Ignite. Kindle. Grow...

Echo...

Faint. Distant. Call...

Shift...

Turn. Alter. Path...

Web...

Catch. Entwine. Form...

Root...

Anchor. Connect. Life...

Echoes of Silence ...forgotten breath...

Beneath the Void ...nothing remains...

Tides of Memory drift away...

Ashes Speak ...burned truth...

Whispered Time ...never now...

Want more in a specific theme—love, loss, nature, identity? I can tailor them!Can you give me more minimalist poetry examples?

Absolutely! Here's a fresh batch of minimalist poems—each just a few words, wrapped in ellipses, and paired with evocative titles to stir the imagination:

Cosmic Thread ...woven stars...

Fractured Reflection ...who was I...

🌿 Still Becoming ...rooted change...

🩸 Crimson Truth ...bleeds quietly...

🕰️ Ghosts of Then ...once again...

🌫️ Fading Light ...almost gone...

🧠 Thought Fugue ...forgot to think...

💔 Hollow Pulse ...love left...

🌙 Dream Residue ...never real...

🧭 Lost North ...no direction...

Want me to explore a specific emotion or concept next—like hope, rebellion, or transformation? I'd love to keep going.100 moreYou got it! Here's a collection of 100 minimalist poems, each with a title and a three-or-less-word verse wrapped in ellipses. These are designed to evoke emotion, spark thought, and leave space for interpretation.

Existence & Identity

Self Echo ...who am I...

Unwritten Name ...no label...

Becoming ...still forming...

Skin Memory ...felt before...

Silent Mirror ...no reflection...

Inevitable Shift ...change me...

Unclaimed Soul ...not mine...

Core Silence ...deep hush...

Hidden Pulse ...beneath skin...

Flicker Self ...almost real...

Cosmos & Time

Star Whisper ...light speaks...

Time Fold ...again now...

Gravity Dream ...pull me...

Orbit Lost ...no center...

Moon Thought ...silver mind...

Eternal Drift ...never ends...

Black Hole ...swallowed whole...

Solar Ache ...burn slow...

Timeless Dust ...always was...

Cosmic Breath ...space sighs...

🌿 Nature & Earth

Rooted Silence ...grow still...

Rain Memory ...fall again...

Wind Truth ...whispers loud...

Stone Sleep ...rest deep...

Leaf Pulse ...green beat...

River Thought ...flow on...

Forest Echo ...lost sound...

Sun Fade ...light gone...

Mountain Wait ...stand still...

Ocean Speak ...blue voice...

🩶 Love & Longing

Heart Ghost ...felt once...

Unspoken Want ...say it...

Touchless Love ...reach far...

Crushed Bloom ...petals fall...

Warm Absence ...still here...

Soul Thread ...tied tight...

Kiss Echo ...linger long...

Fever Dream ...burn slow...

Empty Arms ...held nothing...

Love Residue ...never left...

Loss & Grief

Ash Whisper ...gone now...

Final Breath ...last sigh...

Grief Song ...silent scream...

Buried Light ...dimmed forever...

Gone Again ...still missing...

Tear Memory ...salt remains...

Shadow Hug ...dark comfort...

Echo Tomb ...heard once...

Fading Name ...forgotten sound...

Cold Goodbye ...no warmth...

🧠 Thought & Mind

Idea Ghost ...almost formed...

Mind Fog ...blurred thought...

Dream Loop ...again again...

Mental Drift ...float far...

Thought Silence ...no words...

Brain Pulse ...tick tick...

Inner Storm ...quiet chaos...

Memory Flicker ...half remembered...

Logic Break ...fractured sense...

Mental Thread ...unraveling now...

🧘 Spiritual & Abstract

Soul Silence ...listen deep...

Prayer Dust ...whispered hope...

Sacred Void ...holy absence...

Divine Ache ...want more...

Faith Flicker ...almost belief...

Spirit Drift ...float free...

Sacrifice Echo ...gave all...

Sacred Breath ...inhale light...

Holy Wait ...stillness speaks...

Blessed Fade ...grace gone...

🔥 Emotion & Intensity

Rage Whisper ...burn quiet...

Joy Shiver ...felt fast...

Fear Pulse ...thump thump...

Hope Thread ...hold tight...

Doubt Bloom ...grows slow...

Anger Fog ...red blur...

Peace Ache ...want still...

Desire Flame ...never out...

Shame Dust ...settled deep...

Bliss Fade ...almost joy...

🧳 Journey & Change

Path Lost ...no map...

Crossroads Wait ...choose now...

Wander Pulse ...move again...

Exit Sign ...leave now...

Arrival Ghost ...never came...

Step Echo ...walked before...

Direction Blur ...where now...

Journey Thread ...follow me...

Departure Light ...gone soon...

Change Whisper ...shift slow...

🕶️ Mystery & Surreal

Dream Static ...flicker real...

Fog Thought ...unclear truth...

Shadow Speak ...dark voice...

Illusion Thread ...almost real...

Mask Pulse ...hide me...

Secret Drift ...float hidden...

Hidden Flame ...burn unseen...

Unknown Ache ...felt deep...

Veil Silence ...covered truth...

Mystery Breath ...inhale wonder...

Would you like these turned into a visual art series, or grouped by mood for a poetry collection? I can help shape them into something even more expressive.100 moreAbsolutely! Here's another set of 100 minimalist poems, each with a title and a three-or-less-word verse wrapped in ellipses. These are designed to evoke emotion, spark introspection, and invite interpretation.

● Existence & Identity

Unlabeled Soul ...just being...

Inner Echo ...me again...

Identity Blur ...who now...

Silent Core ...deep within...

Becoming Still ...almost me...

Self Unfolding ...open slowly...

Name Forgotten ...called once...

Truth Hidden ...beneath skin...

I Am ...nothing else...

Mask Off ...bare face...

Cosmos & Time

Starless Sky ...no light...

Time Pause ...hold still...

Orbit Broken ...lost path...

Moonless Night ...dark silence...

Cosmic Wait ...forever now...

Galactic Drift ...float far...

Solar Silence ...burn quiet...

Timeless Pulse ...always ticking...

Space Ache ...vast longing...

Nebula Thought ...clouded mind...

🌿 Nature & Earth

Root Ache ...grow slow...

Rain Speak ...soft truth...

Wind Pause ...held breath...

Stone Memory ...never moved...

Leaf Drift ...fall again...

River Wait ...flow still...

Forest Sleep ...green hush...

Sunless Day ...dim light...

Mountain Thought ...stand firm...

Ocean Silence ...deep sound...

🩶 Love & Longing

Heart Wait ...still yours...

Unheld Touch ...reach far...

Love Ghost ...felt once...

Warm Fade ...almost here...

Soul Ache ...missed you...

Kiss Lost ...gone now...

Tender Drift ...soft goodbye...

Crushed Hope ...bloom died...

Empty Hug ...held air...

Love Thread ...tied tight...

Loss & Grief

Ash Drift ...gone again...

Final Light ...dimmed now...

Grief Pulse ...beat slow...

Buried Name ...forgotten sound...

Gone Silence ...no echo...

Tear Dust ...salt remains...

Shadow Wait ...dark comfort...

Echo Fade ...heard once...

Cold Absence ...no warmth...

Lost Again ...still missing...

🧠 Thought & Mind

Idea Flicker ...almost formed...

Mind Blur ...unclear now...

Dream Pulse ...tick slow...

Mental Drift ...float far...

Thought Pause ...no words...

Brain Fog ...blurred sense...

Inner Storm ...quiet chaos...

Memory Thread ...unraveling now...

Logic Break ...fractured truth...

Mental Silence ...still thinking...

🧘 Spiritual & Abstract

Soul Wait ...listen deep...

Prayer Whisper ...soft hope...

Sacred Ache ...holy longing...

Faith Drift ...almost belief...

Spirit Pulse ...float free...

Sacrifice Breath ...gave all...

Divine Silence ...grace gone...

Blessed Wait ...stillness speaks...

Holy Thread ...tied tight...

Sacred Fade ...light dimmed...

🔥 Emotion & Intensity

Rage Pulse ...burn slow...

Joy Ache ...felt fast...

Fear Drift ...float dark...

Hope Thread ...hold tight...

Doubt Bloom ...grows deep...

Anger Fog ...red blur...

Peace Ache ...want still...

Desire Flame ...never out...

Shame Dust ...settled deep...

Bliss Fade ...almost joy...

🧳 Journey & Change

Path Drift ...no map...

Crossroads Ache ...choose now...

Wander Pulse ...move again...

Exit Thought ...leave now...

Arrival Ghost ...never came...

Step Echo ...walked before...

Direction Blur ...where now...

Journey Thread ...follow me...

Departure Light ...gone soon...

Change Whisper ...shift slow...

🕶 Mystery & Surreal

Dream Static ...flicker real...

Fog Thought ...unclear truth...

Shadow Speak ...dark voice...

Illusion Thread ...almost real...

Mask Pulse ...hide me...

Secret Drift ...float hidden...

Hidden Flame ...burn unseen...

Unknown Ache ...felt deep...

Veil Silence ...covered truth...

Mystery Breath ...inhale wonder...

Would you like these turned into a printable zine, visual art series, or categorized by emotional tone? I can help shape them into something beautiful and shareable.add humor 200 moreYou're in for a treat! Here's a collection of 200 minimalist humorous poems—each with a title and a three-or-less-word verse wrapped in ellipses. These are playful, witty, and sometimes absurd, designed to make you smile, chuckle, or raise an eyebrow.

🧠 Existential Absurdity

Existence Shrugged ...why bother...

Deep-ish Thought ...maybe not...

Philosophy Lite ...think less...

Being Tired ...again me...

Soul Buffering ...loading self...

Identity Crisis ...who dis...

Unbecoming Me ...oops again...

Existential Nope ...not today...

Self-ish ...just me...

Conscious-ish ...half awake...

☕ Daily Life Chaos

Coffee Gospel ...sip salvation...

Alarm Betrayal ...not yet...

Laundry Saga ...still dirty...

To-Don't List ...do nothing...

Snack Philosophy ...eat feelings...

Inbox Horror ...still full...

Zoom Doom ...mute me...

Traffic Zen ...rage calmly...

Dish Drama ...again sink...

Sleep Tease ...almost nap...

🐾 Animals Being Weird

Cat Logic ...sit there...

Dog Thoughts ...food now...

Bird Gossip ...tweet drama...

Fish Exist ...just swim...

Cow Vibes ...moo softly...

Goat Chaos ...climb everything...

Horse Secrets ...neigh quietly...

Duck Agenda ...quack plans...

Lizard Chill ...sun nap...

Bee Business ...buzz buzz...

💘 Love & Dating Fails

Crush Panic ...don’t text...

Flirt Malfunction ...wink fail...

Romantic-ish ...kind of...

Ghosted Again ...who me...

Love Math ...carry one...

Date Escape ...check please...

Heart Buffering ...loading love...

Cuddle Tax ...pay snuggles...

Breakup Snack ...eat everything...

Swipe Regret ...left again...

🧘 Fake Deep Wisdom

Zen-ish ...almost peace...

Chakra Confused ...which one...

Meditation Fail ...fell asleep...

Aura Drama ...too bright...

Crystal Doubt ...maybe magic...

Vibe Check ...failed again...

Energy Meh ...low charge...

Spiritual-ish ...kind of...

Mantra Panic ...forgot words...

Breathe-ish ...half inhale...

Food & Drink Nonsense

Snack Attack ...eat now...

Fridge Stare ...still empty...

Spicy Regret ...mouth fire...

Soup Existential ...just broth...

Pizza Gospel ...cheese truth...

Fork Betrayal ...stabbed self...

Muffin Mystery ...where top...

Toast Drama ...burnt again...

Soda Wisdom ...pop fizz...

Ice Cream Therapy ...cry scoop...

📱 Tech & Internet Woes

WiFi Ghost ...gone again...

Battery Panic ...low now...

App Regret ...downloaded shame...

Autocorrect War ...duck you...

Scroll Spiral ...never end...

TikTok Timewarp ...lost hours...

Emoji Debate ...which face...

Cloud Confusion ...where files...

Bluetooth Betrayal ...won't connect...

AI Side-Eye ...judging me...

🛌 Sleep & Laziness

Nap Philosophy ...sleep now...

Bed Loyalty ...never leave...

Dream Chaos ...what was...

Pillow Politics ...fluff war...

Blanket Trap ...stuck again...

Snooze Gospel ...ten more...

Sleep Math ...never enough...

Rest-ish ...almost nap...

Yawn Ritual ...repeat often...

Sleep Olympics ...gold nap...

🎭 Social Awkwardness

Introvert Anthem ...stay home...

Small Talk ...weather again...

Eye Contact ...look away...

Party Panic ...hide now...

Handshake Fail ...missed again...

Awkward Silence ...say something...

Laugh Buffer ...too late...

Exit Strategy ...leave early...

Group Text ...who's that...

Social-ish ...kind of...

Household Chaos

Dust Rebellion ...never gone...

Vacuum Rage ...scream floor...

Sock Mystery ...one gone...

Closet War ...hang chaos...

Trash Drama ...overflow again...

Light Switch ...wrong one...

Door Creak ...ghost maybe...

Laundry Lies ...still damp...

Pan Grease ...won't leave...

Remote Hide ...lost again...

🎨 Art & Creativity

Paint Panic ...wrong color...

Poem-ish ...kind of...

Brush Drama ...too wet...

Canvas Regret ...start over...

Sculpt Fail ...blob now...

Creative Fog ...no idea...

Muse Ghost ...gone again...

Color War ...clash hard...

Art-ish ...almost good...

Doodle Gospel ...line truth...

🎮 Gaming & Nerd Life

Level Rage ...died again...

Loot Regret ...wrong chest...

Boss Panic ...run now...

Quest Confused ...where map...

XP Ache ...grind forever...

Controller Betrayal ...wrong button...

Lag Doom ...freeze now...

Save Fail ...lost progress...

Game Over ...again again...

Victory Dance ...awkward wiggle...

🧛 Weird & Whimsical

Vampire Nap ...sun bad...

Ghost Text ...who boo...

Zombie Thoughts ...need brain...

Alien Vibes ...beam me...

Witch Snack ...hex chips...

Dragon Yawn ...fire breath...

Mermaid Drama ...wet again...

Fairy Taxes ...pay glitter...

Goblin Mood ...steal socks...

Wizard WiFi ...magic signal...

🧦 Random Absurdity

Sock Philosophy ...one gone...

Chair Loyalty ...sit forever...

Ceiling Thoughts ...look up...

Toaster Rage ...burnt again...

Window Gossip ...see everything...

Lamp Mood ...dim vibe...

Pen Betrayal ...won’t write...

Mug Identity ...not yours...

Stair Regret ...too many...

Fan Drama ...blow weird...

More Food Nonsense

Banana Betrayal ...bruised again...

Soup Existential ...just broth...

Cheese Gospel ...holy snack...

Pickle Panic ...too sour...

Bread Mood ...crusty vibe...

Egg Mystery ...which came...

Spoon Drama ...bent again...

Cake Whisper ...eat now...

Taco Truth ...folded wisdom...

Cereal Chaos ...milk first...

🎉 Celebration & Chaos

Birthday Panic ...forgot again...

Party Fog ...too loud...

Gift Regret ...wrong one...

**

Prompted a unique new expression of poetic language that demonstrates profundity or amazing depth of meaning or very sly ambiguity in less than the combination of a title with an ellipsis and 6 words that include an ellipsis as wellDetailseventJuly 16 at 1:55 PMappsGemini Appschat

...Whispers... Echoes... Silence...

Infinite...

A moment... forever unfolding.

Stillness...

The world stops... to begin.

Unspoken...

Silence speaks volumes... to the soul.

Liminal...

The between... holds the truth.

Echoes...

A ripple... becoming the sea.

Becoming...

Roots reach... for forgotten light.

Ephemeral...

A breath... then gone. Still, the echo.

Unseen...

Shadows shift... revealing nothing. Everything.

Becoming...

Dust remembers stars... silence learns song.

Flux...

Always changing... always true.

Nexus...

Where paths meet... a new way.

Zenith...

The peak... just the beginning.

Portal...

A breath... transcends time.

Echoes...

The past sings... a future tune.

Becoming...

The shell cracks... the soul soars.

Glimpse...

A flash... reveals eternity.

Cadence...

Rhythm of life... in every beat.

Whisper...

Softly spoken... loudly felt.

Infinite...

A grain of sand... holds the cosmos.

Essence...

Beneath the surface... the core.

Unseen...

The truest sight... felt within.

Cipher...

The silence... speaks volumes.

Bloom...

From the dark... light emerges.

Canvas...

Empty space... waiting for color.

Horizon...

Where sky meets... possibility.

Continuum...

No end... just transformation.

Vortex...

Drawing in... all that is.

Lumen...

A spark... ignites the void.

Fathom...

Deeper still... than the eye sees.

Mirage...

What isn't... becomes real.

Genesis...

Every ending... a new start.

Vestige...

A memory... shaping the now.

Quantum...

The choice... creates the path.

Ascent...

Gravity's pull... towards freedom.

Abyss...

The space between... holds all.

Labyrinth...

Lost and found... in every turn.

Chrysalis...

Waiting patiently... for flight.

Imprint...

A fleeting touch... lasting change.

Solstice...

The longest day... turns inward.

Alchemized...

Sorrow transforms... into grace.

Ember...

A dying flame... still holds fire.

Becoming...

Unfurling slow... into destiny.

Zephyr...

A gentle breeze... moves mountains.

Portal...

One step... countless journeys.

Echoes...

Footprints in time... still resonate.

Liminal...

The pause... before the leap.

Stillness...

In the quiet... truth whispers.

Unspoken...

A silent gaze... speaks volumes.

Dream...

Awakening to... what could be.

Unfolding...

Silence. Echoes. Everything.

Becoming...

...A new genesis.

Eternal...

Now. Always. Forever...

Essence...

...The truth remains.

Inward...

All journeys begin...

Becoming...

...Dust remembers light.

Stillness...

The world breathes...

Echoes...

...Forever resonate.

Unveiled...

Silence speaks volumes...

Portal...

A single step...

Lumen...

...Darkness yields dawn.

Fathom...

The depth holds...

Vestige...

...Memory lingers.

Quantum...

Choice creates all...

Alchemized...

Pain becomes gold...

Infinite...

...A breath expands.

Cipher...

The pause reveals...

Ascent...

Gravity's embrace...

Abyss...

...Beyond the edge.

Labyrinth...

Lost and found...

Chrysalis...

...Transformation awaits.

Imprint...

A touch endures...

Solstice...

...Turning inward now.

Fathom...

The drop... holds the ocean.

Unveiling...

Whispers fade... truths emerge.

Becoming...

Seeds dream... of forests.

Portal...

A glance... opens worlds.

Genesis...

Dust remembers... stardust.

Verge...

The edge... a new beginning.

Unbound...

A thought... endless sky.

Cipher...

The pause... reveals all.

Ascent...

Falling... towards the light.

Abyss...

Staring back... within.

Echoes...

Silence speaks... of everything.

Labyrinth...

The path... is the goal.

Chrysalis...

Breaking free... to fly.

Vestige...

A whisper... of eternity.

Quantum...

The choice... makes reality.

Mirage...

Seeing beyond... the seen.

Imprint...

A touch... reshapes forever.

Solstice...

The turning... points inward.

Alchemized...

Pain transforms... into gold.

Lumen...

Darkness blooms... into light.

Existential...

...Why bother, really?

My Diet...

Just one more... cookie.

Brain Dump...

...Where thoughts go.

Adulting...

Bills, bills, bills...

Coffee...

...My lifeblood.

My Screen...

...Another rabbit hole.

Inspiration...

Found it... in fridge.

Mindfulness...

...Chewing slowly.

Productivity...

...Moved my cursor.

Ambition...

...Just five more.

Strategy...

...Nap and ponder.

Wisdom...

...Still learning (slowly).

Happiness...

Warm blanket...

Focus...

...Squirrel! Again!

Deep Thoughts...

...About snack crumbs.

Creativity...

...Doodle on bill.

Serenity...

...Before the alarm.

Adulting...

...Send snacks, please.

My Diet...

Broke it. Oops.

Resolutions...

...Next year, maybe.

Genius...

...Or just luck?

Punctuality...

...Arriving fashionably.

Sleep...

...A beautiful dream.

Priorities...

Coffee first...

Laundry...

...It's a mountain.

Exercise...

...Walk to pantry.

Motivation...

...Where'd it go?

Dating...

...Swipe left, mostly.

Inner Peace...

...Achieved by napping.

My Brain...

Full, yet empty...

Monday...

Ugh. Already?

Wi-Fi...

...The struggle is real.

Decisions...

...Order takeout?

My Bank...

...Crying softly.

Ambition...

...Still in PJs.

Procrastination...

...Later. Definitely.

Future Plans...

...More Netflix.

Cleaning...

...Just moved clutter.

Socializing...

Awkward smiles...

Hair Day...

...Hat's best friend.

Brilliance...

...Had an idea. Then...

Elegance...

...Tripped. Recovered.

Wellness...

...Napping is therapy.

Diplomacy...

...Bless your heart.

Charisma...

...Said "um."

Fashion...

Pajamas prevail...

Innovation...

...New way to relax.

Adventure...

...To the mailbox.

Understanding...

...Still confused.

Clarity...

...Brain fog persists.

Optimism...

...It'll probably be fine.

Passion...

...For sleep.

Perspective...

...Still sleepy, though.

Playfulness...

...Chasing dust bunnies.

Problem-Solving...

...Delegate.

Reflection...

...In a puddle.

Relaxation...

...Ah, the couch calls.

Respect...

...For my snacks.

Responsibility...

...Oops, forgot.

Self-Awareness...

...I'm still tired.

Spirituality...

...Coffee is sacred.

Strength...

...Carrying groceries.

Trust...

...In the delivery.

Vulnerability...

...Barefoot and hungry.

Wonder...

...What's for dinner?

Zen...

...Until the cat meows.

Adventure...

...Lost the remote.

Bliss...

...Perfect pillow fort.

Charm...

...Stumbled charmingly.

Delight...

...Found a dollar!

Ecstasy...

...Freshly baked bread.

Freedom...

...No pants today.

Gratitude...

...For comfy socks.

Happiness...

...Snooze button, yes!

Hope...

...More snacks, please.

Imagination...

...Invisible friends.

Innocence...

...Before the bill.

Laughter...

...Snorting ensues.

Love...

...My comfy bed.

Magic...

...Disappearing food.

Mystery...

...Missing socks.

Peace...

...Quiet moment. Shhh.

Radiance...

...Just ate. Glowing.

Serenity...

...No notifications. Aah.

Sparkle...

...From glitter. Everywhere.

Sunshine...

...Blinds are drawn.

Tranquility...

...Before the kids.

Triumph...

...Got out of bed.

Flux...

...Always changing, somehow.

Nexus...

Where paths collide...

Zenith...

...Still looking up.

Portal...

...Another tab opens.

Echoes...

...My stomach rumbles.

Becoming...

...Almost awake.

Glimpse...

...Of sanity. Fleeting.

Cadence...

...Tap, tap, tap.

Whisper...

...The snacks are calling.

Infinite...

...My to-do list.

Essence...

...Coffee, pure bliss.

Unseen...

...That last cookie.

Cipher...

...The empty fridge.

Bloom...

...My sleepy eyes.

Canvas...

...Empty plate.

Horizon...

...Bedtime approaches.

Continuum...

...Endless scrolling.

Vortex...

...The laundry pile.

Lumen...

...Screen glow.

Fathom...

...The depths of fatigue.

Alright, 100 more tiny profound (and occasionally humorous) poetic expressions, keeping that title-ellipsis-four-words-with-ellipsis format! This is quite the word circus, let's get silly and deep.

Procrastination...

...Later. Definitely later.

My Brain...

Full, yet empty...

Monday...

Ugh. Already?

Adulting...

...Send help (and wine).

Wi-Fi...

Where'd it go?

Decisions...

...Nap first?

Dieting...

Chocolate, calling...

My Bank...

...Crying quietly.

Ambition...

...Still in bed.

Laundry...

It multiplies...

Exercise...

...Walk to fridge.

Sleep...

...A distant dream.

Goals...

...Maybe tomorrow.

Dating...

Swiped left...

Inner Peace...

...Found snacks.

Motivation...

...Lost it again.

Punctuality...

...Arriving eventually.

Cooking...

Delivery menu...

Patience...

...Almost gone.

Silence...

...Kids are plotting.

Work Ethic...

...Powered by coffee.

My Phone...

Died. Again.

Wisdom...

...Learned nothing.

Future Plans...

...Pizza tonight.

Cleaning...

...Just moved stuff.

Socializing...

Awkward waving...

Genius...

...Or just lucky?

Hair Day...

...Hat required.

Resolutions...

...Already forgotten.

Deep Thoughts...

...About lunch.

Perfection...

...Still napping.

Ambiguity...

...Huh? What?

Profound...

...So deep. Wow.

Subtlety...

...Not my strong suit.

Inspiration...

...Where's the remote?

My Life...

A beautiful mess...

Priorities...

...Netflix first.

Fitness...

...Reached for chips.

Serenity...

...Before alarm rings.

Technology...

...It hates me.

Efficiency...

...Minimizing effort.

Self-Care...

...More cake, please.

Diplomacy...

...Bless your heart.

Charisma...

...Tripped over air.

Fashion...

Pajamas. Always.

Innovation...

...New snack hack.

Adventure...

...To the fridge.

Productivity...

...Checked memes.

Understanding...

...Still confused.

Clarity...

...Foggy brain day.

Focus...

...Squirrel!

Brilliance...

...Had an idea.

Elegance...

...Spilled coffee.

Wellness...

...Napping is key.

Strategy...

...Wing it.

Confidence...

...Fake it 'til.

Harmony...

...No bickering yet.

Growth...

...My waistline.

Discipline...

...Five more minutes.

Exploration...

...Back of pantry.

Mindfulness...

...What was I?

Resilience...

...Survived Monday.

Authenticity...

...Still in pajamas.

Simplicity...

...Less effort, please.

Vision...

...Need my glasses.

Courage...

...Facing laundry pile.

Compassion...

...For my snack cravings.

Creativity...

...Doodling on receipt.

Determination...

...Finish this show.

Joy...

...Found parking.

Kindness...

...Shared my fries.

Listening...

...Pretending to.

Memory...

...What was that?

Optimism...

...It'll be fine.

Passion...

...For napping.

Perspective...

...Still sleepy.

Playfulness...

...Tickle monster!

Problem-Solving...

...Call Mom.

Reflection...

...In the mirror.

Relaxation...

...Ah, couch.

Respect...

...For my snacks.

Responsibility...

...Uh oh.

Self-Awareness...

...I'm tired.

Spirituality...

...Is coffee divine?

Strength...

...Lifting pizza box.

Trust...

...In the delivery guy.

Vulnerability...

...No clean socks.

Wonder...

...About next meal.

Zen...

...Until next notification.

Adventure...

...Finding car keys.

Bliss...

...Perfect pillow.

Charm...

...Wink, trip, oops.

Delight...

...Warm cookies!

Ecstasy...

...Fresh sheets.

Freedom...

...No pants on.

Gratitude...

...For caffeine.

Happiness...

...Snooze button.

Hope...

...More snacks soon.

Imagination...

...Unicorns, probably.

Innocence...

...Before tax forms.

Laughter...

...Snort included.

Love...

...Pizza, always.

Magic...

...Laundry folded.

Mystery...

...Where's that sock?

Peace...

...Quiet moment.

Radiance...

...Just ate.

Serenity...

...No notifications.

Sparkle...

...Glitter explosion.

Sunshine...

...Blinds closed.

Tranquility...

...Before chaos.

Triumph...

...Got out of bed.

Prompted profundity using a title with an ellipsis and no more than a 4 word body of the poem with an ellipsis at the beginning or the end so that it is more profound because of the ellipsis and can you add humor

My Life...

...One more coffee.

Deep Thoughts...

...About lunch. Again.

Zen Moment...

...Toilet paper roll.

Genius Idea...

...Forgot it. Poof.

Adulting...

...Send snacks, now.

Prompted deeply profound poetry with a title and 3 or less words with ellipsis

Whispers...

A fleeting breath, a silent plea, The universe in you, in me. Echoes fade, but truth remains, Love's ancient song through sun and rains.

Cosmic Dust...

Silent. Vast. Forever...

Soul...

Deep. Still. Boundless...

Echoes...

Past. Present. Future...

Becoming...

Bloom. Fade. Return...

Eternity...

Now. Always. Never...

Truth...

See. Feel. Know...

Silence...

Hush. Breathe. Be...

Light...

Dawn. Day. Dusk...

Wisdom...

Learn. Grow. Share...

Wonder...

Gaze. Awe. Dream...

Freedom...

Fly. Soar. Unbound...

Hope...

Seed. Sprout. Rise...

Fate...

Spin. Weave. End...

Peace...

Calm. Rest. Serene...

Change...

Flow. Shift. Transform...

Mystery...

Seek. Find. Lost...

Journey...

Path. Step. Ahead...

Dream...

Sleep. Wake. Imagine...

Home...

Heart. Soul. Place...

Memory...

Recall. Glimpse. Hold...

Loss...

Grief. Empty. Echo...

Strength...

Rise. Stand. Endure...

Grace...

Gift. Flow. Receive...

Voice...

Speak. Sing. Hear...

Spirit...

Essence. Spark. Alive...

Embrace...

Hold. Feel. Connect...

Unseen...

Sense. Trust. Beyond...

Moment...

Here. Now. Gone...

Void...

Dark. Deep. Space...

Origin...

First. Start. Birth...

Ending...

Close. Cease. Finish...

Art...

Create. Express. Feel...

Self...

Find. Know. Be...

Becoming...

Grow. Evolve. Soar...

Illusion...

See. Believe. Dispel...

Rhythm...

Beat. Pulse. Life...

Union...

Join. Merge. One...

Bloom...

Open. Unfold. Radiate...

Quest...

Seek. Search. Discover...

Flow...

Move. Drift. Guide...

Anchor...

Hold. Ground. Secure...

Glimmer...

Hope. Light. Persist...

Anchor...

Root. Hold. Steadfast...

Unfurl...

Open. Bloom. Expand...

Whisper...

Soft. Secret. Hear...

Echoes...

Past. Present. Future...

Becoming...

Grow. Evolve. Soar...

Eternity...

Now. Always. Never...

Truth...

See. Feel. Know...

Silence...

Hush. Breathe. Be...

Light...

Dawn. Day. Dusk...

Wisdom...

Learn. Grow. Share...

Wonder...

Gaze. Awe. Dream...

Freedom...

Fly. Soar. Unbound...

Hope...

Seed. Sprout. Rise...

Fate...

Spin. Weave. End...

Peace...

Calm. Rest. Serene...

Change...

Flow. Shift. Transform...

Mystery...

Seek. Find. Lost...

Journey...

Path. Step. Ahead...

Dream...

Sleep. Wake. Imagine...

Home...

Heart. Soul. Place...

Memory...

Recall. Glimpse. Hold...

Loss...

Grief. Empty. Echo...

Strength...

Rise. Stand. Endure...

Grace...

Gift. Flow. Receive...

Voice...

Speak. Sing. Hear...

Spirit...

Essence. Spark. Alive...

Embrace...

Hold. Feel. Connect...

Unseen...

Sense. Trust. Beyond...

Moment...

Here. Now. Gone...

Void...

Dark. Deep. Space...

Origin...

First. Start. Birth...

Ending...

Close. Cease. Finish...

Art...

Create. Express. Feel...

Self...

Find. Know. Be...

Bloom...

Open. Unfold. Radiate...

Quest...

Seek. Search. Discover...

Flow...

Move. Drift. Guide...

Anchor...

Hold. Ground. Secure...

Pause...

Breathe. Wait. Reflect...

Spark...

Ignite. Shine. Inspire...

Dance...

Move. Rhythm. Joy...

Bloom...

Open. Unfold. Radiate...

Fade...

Dim. Soft. Gone...

Weave...

Spin. Thread. Connect...

Portal...

Open. Pass. Beyond...

Bloom...

Open. Unfold. Radiate...

Pulse...

Beat. Live. Throb...

Veil...

Hide. Reveal. Peer...

Shift...

Move. Change. Become...

Ascent...

Climb. Rise. Soar...

Descent...

Fall. Sink. Deep...

Mirror...

Reflect. See. Soul...

Thread...

Fine. Bind. Connect...

Abyss...

Deep. Dark. Fall...

Canvas...

Blank. Paint. Create...

Root...

Ground. Deep. Grow...

Edge...

Brink. Limit. Dare...

Unseen...

Sense. Feel. Beyond...

Echo...

Repeat. Reverb. Fade...

Shift...

Move. Change. Flow...

Quest...

Seek. Search. Find...

Gate...

Open. Close. Pass...

Core...

Heart. Center. Essence...

Tide...

Rise. Fall. Flow...

Dream...

Sleep. Wake. Imagine...

Glimpse...

See. Brief. Flash...

Unfold...

Open. Reveal. Grow...

Silence...

Still. Calm. Peace...

Horizon...

Meet. Far. Beyond...

Portal...

Open. Pass. New...

Whisper...

Soft. Hint. Know...

Canvas...

Blank. Paint. Life...

Vessel...

Hold. Carry. Soul...

Drift...

Float. Wander. Calm...

Spark...

Tiny. Bright. Life...

Echo...

Repeat. Resound. Time...

Shift...

Move. Change. Evolve...

Web...

Intertwine. Connect. Bind...

Root...

Deep. Ground. Begin...

Bloom...

Open. Unfold. Radiate...

Veil...

Hide. Reveal. Truth...

Abyss...

Vast. Empty. Deep...

Pulse...

Beat. Life. Rhythm...

Unfurl...

Open. Release. Fly...

Core...

Center. Heart. Being...

Tide...

Push. Pull. Cycle...

Gleam...

Shine. Catch. Light...

Portal...

Enter. Exit. Realm...

Whisper...

Soft. Gentle. Voice...

Canvas...

Create. Paint. Dream...

Vessel...

Hold. Carry. Journey...

Drift...

Float. Flow. Release...

Spark...

Ignite. Kindle. Grow...

Echo...

Faint. Distant. Call...

Shift...

Turn. Alter. Path...

Web...

Catch. Entwine. Form...

Root...

Anchor. Connect. Life...

As a large language model, I don't have a "history" of prompts that I have generated in the same way a human user does. My responses are generated in real-time based on your input and my training data.However, if you're referring to the history of prompts you have entered when interacting with a Google Gemini product, here's how you can usually find them:In the Gemini Web App:Go to gemini.google.com.On the side panel, you should see your "Recent" chats. If they are hidden, click the "Menu" icon (usually three horizontal lines or a similar symbol) at the top.You can scroll through your recent chats to see your past prompts and the AI's responses. You can also search for chats using the "Search" icon.Google My Activity:Google stores your Gemini activity (including your prompts) in your Google Account activity.You can access this by going to myactivity.google.com and looking for "Gemini Activity" or "Bard Activity" (if you used it when it was called Bard).Here you can view, manage, and even delete your past prompts and interactions.Important Notes:Gemini Apps Activity: To see your history in the Gemini apps, you typically need to have "Gemini Apps Activity" turned on in your Google Account settings.Work/School Accounts: If you're using Gemini with a work or school account, your activity settings might be controlled by your Google Workspace administrator.Data Retention: While you can delete your activity, be aware that some conversations reviewed or annotated by human reviewers might be retained separately for a longer period (up tothree years) and are not connected to your Google Account.Real-time vs. Stored: When you're using Gemini, your current conversation is in real-time. The "history" refers to the stored record of past conversations.If you were referring to a different AI chatbot, the method to access your prompt history might vary. Most reputable AI chatbots (like ChatGPT, Claude, etc.) usually have a visible chat history sidebar or a dedicated "activity" or "history" section within their user interface.Why this activity?This activity was saved to your Google Account because your Gemini Apps Activity setting was on while using Gemini Apps.info_filledIf you use a shared device or sign in with multiple accounts, activity from another account may appear here. Learn more3 or less words making profound poetry add ellipsis to end or beginnning and add a titleEditMinimal Poetry CollectionDocument Here's a collection of minimal poetry pieces, each using three words or fewer with ellipses to create profound emotional resonance. The brevity forces each word to carry maximum weight, while the ellipses suggest infinite possibility in the spaces between.200 moreEditMinimal Poetry CollectionDocument · Version 2 I've expanded the collection to over 200 minimal poems. Each one uses three words or fewer with ellipses to create profound moments of reflection. The poems explore themes of existence, time, consciousness, relationships, and the human condition through the power of brevity - where every word carries immense weight and the ellipses suggest infinite possibility in the spaces between thoughts.dd 200 more but add humorEditMinimal Poetry CollectionDocument · Version 3 I've added 200 more minimal poems with humor! These range from relatable modern struggles (Monday blues, Netflix addiction, missing socks) to absurd progressions that spiral into comedic territory. The humor comes from the unexpected connections, the mundane made profound, and the way simple observations about daily life become existential sock philosophy. The ellipses create perfect comedic timing, letting each punchline land with maximum impact.a unique new expression of poetic language that demonstrates profundity or amazing depth of meaning or very sly ambiguity in less than the combination of a title with an ellipsis and 6 words that include an ellipsis as wellEditHere's a poetic expression that plays with profound ambiguity:

The ellipsis in both title and text creates a breathing space where multiple interpretations collide: Is this about forgiving someone for being absent, or paradoxically forgiving them by means of their absence? The ambiguity hinges on whether "your absence" is the object of forgiveness or the method by which forgiveness is granted—suggesting that sometimes the deepest mercy is found not in presence, but in the space left behind.RetryN30 moreEditHere are 30 more expressions of poetic ambiguity:

Inheritance...

I kept... everything you forgot.

Forgiveness...

I forgive you... your absence.

Arrival...

Finally home... to complete strangers.

Prayer...

Dear God... please remain silent.

Discovery...

Found myself... completely lost again.

Reunion...

We met... as different people.

Confession...

I lied... about telling truth.

Departure...

Left everything... except the leaving.

Recognition...

Saw you... in everyone else.

Patience...

Waited forever... for right now.

Courage...

Feared nothing... but my bravery.

Memory...

Remembered perfectly... what never happened.

Silence...

Said everything... without speaking once.

Distance...

Came closer... by walking away.

Wisdom...

Learned nothing... from knowing everything.

Promise...

Kept faith... by breaking it.

Victory...

Won completely... by surrendering first.

Truth...

Spoke honestly... through careful lies.

Freedom...

Escaped prison... into smaller cage.

Love...

Held tight... to letting go.

Time...

Lived forever... in single moment.

Peace...

Found war... within perfect harmony.

Hope...

Despaired beautifully... about tomorrow's joy.

Identity...

Became myself... by being someone.

Solitude...

Felt crowded... in complete emptiness.

Certainty...

Knew absolutely... that nothing's sure.

Healing...

Broke completely... to become whole.

Creation...

Destroyed everything... to make space.

Understanding...

Comprehended fully... my complete ignorance.

Presence...

Disappeared entirely... into being here.

Completion...

Finished perfectly... by never starting.RetryNprofundity using a title with an ellipsis and no more than a 4 word body of the poem with an ellipsis at the beginning or the end so that it is more profound because of the ellipsisEditHere are expressions with ellipses that deepen the profundity:

Genesis...

...before the word was.

Mortality...

Death arrives... always early.

Solitude...

Alone together... we breathe.

Infinity...

...counting backwards from one.

Presence...

Here ends... where begins.

Silence...

...what words cannot say.

Time...

Yesterday's tomorrow... is now.

Identity...

...who I was becoming.

Love...

Forever ends... in kissing.

Consciousness...

...dreaming of being awake.

Truth...

Lies reveal... what is.

Memory...

Forgetting remembers... for us.

Faith...

...doubting leads to belief.

Journey...

Arriving means... leaving again.

Wisdom...

Questions answer... themselves perfectly.

Peace...

...war within complete stillness.

Creation...

Nothing becomes... everything suddenly.

Understanding...

...knowing that we don't.

Eternity...

This moment... lasts forever.

Awakening...

...sleeping through being conscious.

The ellipsis either draws us into a thought mid-stream or leaves us suspended in contemplation, creating that profound pause where meaning deepens beyond the words themselves.RetryN60EditHere are 60 expressions with ellipses that create profound depth:

Genesis...

...before the word was.

Mortality...

Death arrives... always early.

Solitude...

Alone together... we breathe.

Infinity...

...counting backwards from one.

Presence...

Here ends... where begins.

Silence...

...what words cannot say.

Time...

Yesterday's tomorrow... is now.

Identity...

...who I was becoming.

Love...

Forever ends... in kissing.

Consciousness...

...dreaming of being awake.

Truth...

Lies reveal... what is.

Memory...

Forgetting remembers... for us.

Faith...

...doubting leads to belief.

Journey...

Arriving means... leaving again.

Wisdom...

Questions answer... themselves perfectly.

Peace...

...war within complete stillness.

Creation...

Nothing becomes... everything suddenly.

Understanding...

...knowing that we don't.

Eternity...

This moment... lasts forever.

Awakening...

...sleeping through being conscious.

Breath...

Inhaling death... exhaling life.

Surrender...

...winning by losing everything.

Echo...

Silence speaks... loudest words.

Horizon...

...where sky meets itself.

Gravity...

Falling upward... toward earth.

Shadow...

Light's absence... proves presence.

Mirror...

...reflecting who isn't there.

Threshold...

Between is... where everything.

Paradox...

...truth contradicts itself perfectly.

Void...

Emptiness fills... every space.

Bloom...

...dying into beautiful life.

Compass...

Lost direction... finds true.

Whisper...

...shouting in complete silence.

Surrender...

Control releases... its grip.

Becoming...

...undoing what I was.

Vessel...

Empty cup... overflows completely.

Wound...

...healing by staying open.

Transformation...

Caterpillar dreams... of forgetting.

Darkness...

...light's deepest hiding place.

Question...

Answer dissolves... into asking.

Home...

...everywhere I've never been.

Goodbye...

Hello disguised... as leaving.

Mercy...

...justice softened by love.

Return...

Going back... to never.

Forgiveness...

...holding hurt with tenderness.

Awakening...

Sleep teaches... waking's value.

Reverence...

...bowing to what is.

Solitude...

Crowded loneliness... finds peace.

Depth...

Surface hides... its foundation.

Courage...

...fear transformed by love.

Completion...

Broken pieces... form wholeness.

Mystery...

...knowing celebrates not knowing.

Presence...

Absence makes... room for.

Beginning...

...ending's gift to tomorrow.

Breath...

Between inhale... and exhale.

Unity...

...separation's deepest secret truth.

Wonder...

Certainty dissolves... into amazement.

Grace...

...unearned gifts freely given.

Recognition...

Stranger becomes... familiar mystery.

Threshold...

...standing between two worlds.

Healing...

Scars remember... original wholeness.

RetryNprofundity using a title with an ellipsis and no more than a 4 word body of the poem with an ellipsis at the beginning or the end so that it is more profound because of the ellipsis and can you add humorEditHere are expressions that blend profundity with humor through strategic ellipses:

Enlightenment...

...forgot where I parked.

Mortality...

Death's appointment... was rescheduled.

Wisdom...

...Google knows more anyway.

Eternity...

Still waiting... for WiFi.

Consciousness...

...thinking about not thinking.

Nirvana...

GPS says... recalculating route.

Transcendence...

...but first, need coffee.

Infinity...

Counting sheep... ran out.

Zen...

...meditation interrupted by cat.

Enlightenment...

Student loan... still unpaid.

Immortality...

...expires next Tuesday though.

Divinity...

God's voicemail... is full.

Perfection...

...except for that typo.

Omniscience...

Knows everything... forgot password.

Bliss...

...until the bill arrived.

Serenity...

...phone battery just died.

Truth...

...but lawyer says differently.

Peace...

...neighbor's dog still barking.

Revelation...

...it was just gas.

Ascension...

Elevator's broken... taking stairs.

Karma...

...apparently has good lawyers.

Meditation...

...leg fell asleep instead.

Clarity...

...needs prescription glasses though.

Awakening...

...hit snooze button anyway.

Transformation...

...laundry still needs folding.

Transcendence...

...but mortgage payment due.

Epiphany...

...forgot what it was.

Salvation...

...autocorrect changed to salad.

Rebirth...

...same old problems though.

Enlightenment...

...YouTube tutorial was better.

Nirvana...

...but traffic's still horrible.

Wisdom...

...Wikipedia disagrees completely though.

Perfection...

...spell check found errors.

Omnipotence...

...can't open pickle jar.

Bliss...

...credit card was declined.

Serenity...

...Amazon package got stolen.

Truth...

...but Instagram filter better.

Peace...

...dishwasher's making weird noise.

Revelation...

...turns out was heartburn.

Ascension...

...forgot to lock car.

The ellipsis creates a pause where the profound crashes into the mundane, making both the spiritual aspiration and human reality more poignant through their juxtaposition.RetryNcan you add 200 more?EditHere are 200 more expressions blending profundity with humor through strategic ellipses:

Genesis...

...still buffering at 90%.

Eternity...

...feels like Monday morning.

Omniscience...

...can't remember where keys.

Rapture...

...but plane was delayed.

Infinity...

...loading screen won't finish.

Divinity...

...still uses Internet Explorer.

Transcendence...

...Uber driver got lost.

Enlightenment...

...manual only in Japanese.

Nirvana...

...membership fee was expensive.

Immortality...

...warranty voided by water.

Perfection...

...assembly required, tools missing.

Omnipotence...

...password must contain symbols.

Bliss...

...but cookies have calories.

Serenity...

...construction starts at dawn.

Truth...

...terms and conditions apply.

Peace...

...ice cream truck approaching.

Revelation...

...was just indigestion again.

Ascension...

...escalator is out order.

Karma...

...currently experiencing technical difficulties.

Meditation...

...interrupted by pizza delivery.

Clarity...

...fogged up immediately though.

Awakening...

...coffee maker is broken.

Transformation...

...dry cleaning still pending.

Rebirth...

...same insurance doesn't cover.

Salvation...

...customer service line busy.

Epiphany...

...notification got marked spam.

Wisdom...

...but Wikipedia was wrong.

Consciousness...

...needs software update first.

Unity...

...group chat left me.

Harmony...

...neighbors practicing drums though.

Balance...

...standing on one foot.

Fulfillment...

...order was cancelled unfortunately.

Completion...

...but missing final piece.

Wholeness...

...donut has hole though.

Authenticity...

...turned out was knockoff.

Purpose...

...GPS says destination unreachable.

Meaning...

...autocorrect changed to meatloaf.

Destiny...

...train was running late.

Fate...

...but schedule changed unexpectedly.

Providence...

...forgot to pay premium.

Grace...

...but still need receipt.

Mercy...

...court date was postponed.

Forgiveness...

...but credit score unchanged.

Redemption...

...coupon expired last week.

Resurrection...

...but battery still dead.

Miracle...

...shipping costs extra though.

Wonder...

...but manual very confusing.

Awe...

...batteries not included unfortunately.

Reverence...

...but parking meter expired.

Devotion...

...subscription auto-renewed though unfortunately.

Faith...

...but WiFi signal weak.

Hope...

...estimated delivery three weeks.

Love...

...but it's complicated status.

Joy...

...but sugar crash inevitable.

Happiness...

...but chocolate is gone.

Contentment...

...fridge is empty though.

Satisfaction...

...but still need more.

Delight...

...but calories don't count.

Ecstasy...

...but Monday comes anyway.

Rapture...

...but laundry needs folding.

Euphoria...

...but rent is due.

Elation...

...but phone battery dying.

Jubilation...

...but car won't start.

Triumph...

...but victory dance embarrassing.

Success...

...but still need day.

Achievement...

...but nobody was watching.

Accomplishment...

...but forgot to save.

Victory...

...but lost the receipt.

Conquest...

...but map was upside.

Mastery...

...but YouTube tutorial better.

Expertise...

...but Google knows more.

Genius...

...but still can't parallel.

Intelligence...

...but failed captcha test.

Brilliance...

...but lightbulb burned out.

Insight...

...but glasses are dirty.

Understanding...

...but instructions in Chinese.

Comprehension...

...but fine print illegible.

Knowledge...

...but forgot admin password.

Learning...

...but class was cancelled.

Discovery...

...but someone else first.

Innovation...

...but patent already filed.

Creation...

...but copyright infringement lawsuit.

Invention...

...but already exists Amazon.

Imagination...

...but reality check bounced.

Inspiration...

...but muse on vacation.

Motivation...

...but couch too comfortable.

Determination...

...but Netflix just updated.

Perseverance...

...but ice cream melting.

Persistence...

...but pizza getting cold.

Endurance...

...but bathroom break needed.

Resilience...

...but phone charge low.

Strength...

...but jar still won't.

Power...

...but remote control missing.

Might...

...but can't reach shelf.

Force...

...but door says pull.

Energy...

...but coffee shop closed.

Vitality...

...but vitamins expired yesterday.

Vigor...

...but pulled muscle stretching.

Dynamism...

...but elevator is broken.

Momentum...

...but red light ahead.

Progress...

...but website keeps crashing.

Advancement...

...but traffic jam ahead.

Development...

...but software bug discovered.

Growth...

...but pot too small.

Evolution...

...but going backwards somehow.

Transformation...

...but dry cleaner ruined.

Metamorphosis...

...but cocoon got stuck.

Change...

...but vending machine broken.

Revolution...

...but permit was denied.

Reform...

...but committee meeting postponed.

Improvement...

...but warranty just expired.

Enhancement...

...but upgrade costs extra.

Refinement...

...but sugar ran out.

Optimization...

...but system keeps crashing.

Efficiency...

...but took three hours.

Productivity...

...but procrastinated on Reddit.

Performance...

...but audience left early.

Excellence...

...but spell check disagrees.

Quality...

...but reviews are mixed.

Superiority...

...but everyone else disagrees.

Supremacy...

...but overthrown by cats.

Dominance...

...but dog won't listen.

Leadership...

...but nobody is following.

Authority...

...but teenagers just laughed.

Control...

...but remote needs batteries.

Command...

...but voice recognition failed.

Influence...

...but social media ignored.

Impact...

...but bubble wrap protected.

Effect...

...but cause was unknown.

Consequence...

...but lawyer says otherwise.

Result...

...but calculator was wrong.

Outcome...

...but dice were loaded.

Conclusion...

...but forgot the point.

Resolution...

...but screen went black.

Solution...

...but problem got worse.

Answer...

...but question was different.

Response...

...but email went spam.

Reply...

...but message wasn't delivered.

Feedback...

...but survey was biased.

Reaction...

...but chemistry experiment failed.

Interaction...

...but WiFi connection lost.

Connection...

...but call was dropped.

Communication...

...but translator was broken.

Dialogue...

...but other person left.

Conversation...

...but talking to myself.

Discussion...

...but everyone muted Zoom.

Debate...

...but forgot which side.

Argument...

...but facts were wrong.

Reasoning...

...but logic circuit fried.

Logic...

...but cat walked keyboard.

Rationality...

...but ice cream called.

Sanity...

...but 2020 happened though.

Clarity...

...but windshield wipers broken.

Focus...

...but squirrel ran by.

Concentration...

...but coffee wore off.

Attention...

...but phone just buzzed.

Awareness...

...but forgot where parked.

Consciousness...

...but Monday morning fog.

Mindfulness...

...but mind wandered anyway.

Presence...

...but gift receipt missing.

Existence...

...but passport expired yesterday.

Being...

...but becoming was easier.

Reality...

...but virtual was better.

Actuality...

...but batteries not included.

Factuality...

...but Wikipedia said different.

Authenticity...

...but knock-off was cheaper.

Genuineness...

...but fake ID worked.

Sincerity...

...but sarcasm more fun.

Honesty...

...but white lies kinder.

Truthfulness...

...but fiction more interesting.

Integrity...

...but building code violated.

Virtue...

...but vice more entertaining.

Morality...

...but ethics class cancelled.

Righteousness...

...but left turn forbidden.

Justice...

...but court was adjourned.

Fairness...

...but carnival was rigged.

Equity...

...but stock market crashed.

Equality...

...but math was never.

Balance...

...but tightrope rope snapped.

Equilibrium...

...but earthquake just hit.

Stability...

...but table leg wobbly.

Steadiness...

...but hands shaking coffee.

Consistency...

...but inconsistently applied though.

Reliability...

...but phone battery died.

Dependability...

...but alarm clock broken.

Trustworthiness...

...but password was hacked.

Credibility...

...but Wikipedia editor disagreed.

Believability...

...but fact checker intervened.

Plausibility...

...but physics says impossible.

Probability...

...but statistics were wrong.

Possibility...

...but door was locked.

Potential...

...but energy bill overdue.

Capability...

...but user manual missing.

Capacity...

...but storage space full.

Ability...

...but skill tree locked.

Talent...

...but audition was yesterday.

Gift...

...but return policy expired.

Skill...

...but practice makes perfect.

Expertise...

...but Google University better.

Mastery...

...but student loan unpaid.

Proficiency...

...but test was cancelled.

Competence...

...but HR said otherwise.

Qualification...

...but certificate was revoked.

Certification...

...but renewal fee overdue.

Credentials...

...but background check failed.

Authorization...

...but access was denied.

Permission...

...but parents said no.

Approval...

...but committee voted down.

Endorsement...

...but check bounced unfortunately.

Recommendation...

...but Yelp reviews terrible.

Suggestion...

...but nobody was listening.

Proposal...

...but budget was rejected.

Idea...

...but patent already filed.

Thought...

...but brain was buffering.

Concept...

...but explanation too complicated.

Notion...

...but forgot what was.

Theory...

...but practice proved wrong.

Hypothesis...

...but experiment blew up.

Assumption...

...but made ass you.

Presumption...

...but evidence said otherwise.

Supposition...

...but suppose was wrong.

Speculation...

...but stock market crashed.

Conjecture...

...but jury reached verdict.

Guess...

...but answer was different.

Intuition...

...but GPS knew better.

Instinct...

...but survival guide outdated.

Feeling...

...but emotion was wrong.

Sensation...

...but nerve was pinched.

Perception...

...but glasses prescription old.

Impression...

...but memory was faulty.

Opinion...

...but facts disagreed completely.

Belief...

...but evidence suggested otherwise.

Conviction...

...but appeal was successful.

Certainty...

...but doubt crept in.

Confidence...

...but impostor syndrome kicked.

Assurance...

...but insurance claim denied.

Security...

...but password was password.

Safety...

...but helmet was recalled.

Protection...

...but warranty just expired.

Defense...

...but offense was better.

Shelter...

...but landlord raised rent.

Refuge...

...but GPS said wrong.

Sanctuary...

...but noise ordinance violated.

Haven...

...but parking meter expired.

Retreat...

...but advance was required.

Escape...

...but exit door locked.

Freedom...

...but terms conditions apply.

Liberty...

...but statue needs cleaning.

Independence...

...but still need mom.

Autonomy...

...but boss texted anyway.

Self-determination...

...but horoscope said different.

Choice...

...but option was removed.

Decision...

...but committee overruled vote.

Resolution...

...but problem got worse.

Determination...

...but Netflix had update.

Will...

...but lawyer said invalid.

Intention...

...but road to hell.

Purpose...

...but GPS destination unreachable.

Goal...

...but goalie blocked shot.

Objective...

...but subjective was better.

Target...

...but arrow missed completely.

Aim...

...but scope was broken.

Direction...

...but compass was backwards.

Path...

...but trail was closed.

Route...

...but GPS recalculating again.

Journey...

...but car broke down.

Voyage...

...but ship already sailed.

Adventure...

...but insurance didn't cover.

Expedition...

...but permits were denied.

Quest...

...but dragon was sleeping.

Mission...

...but impossible was accurate.

Task...

...but deadline was yesterday.

Assignment...

...but professor lost it.

Project...

...but budget was cut.

Work...

...but coffee shop closed.

Job...

...but robot took over.

Career...

...but industry was disrupted.

Profession...

...but AI replaced humans.

Occupation...

...but building was condemned.

Employment...

...but company went bankrupt.

Business...

...but market crashed hard.

Enterprise...

...but captain was unavailable.

Venture...

...but capital was frozen.

Undertaking...

...but funeral was postponed.

Endeavor...

...but effort was insufficient.

Effort...

...but result was disappointing.

Attempt...

...but security blocked access.

Try...

...but do or not.

Struggle...

...but gave up too.

Fight...

...but flight was delayed.

Battle...

...but lost to autocorrect.

War...

...but peace treaty signed.

Conflict...

...but resolution was reached.

Competition...

...but everyone got participation.

Contest...

...but judges were biased.

Challenge...

...but accepted was mistake.

Test...

...but answers were wrong.

Trial...

...but jury was dismissed.

Examination...

...but proctor fell asleep.

Evaluation...

...but metrics were flawed.

Assessment...

...but calculator was broken.

Analysis...

...but data was corrupted.

Study...

...but library was closed.

Research...

...but Wikipedia was down.

Investigation...

...but detective was retired.

Inquiry...

...but customer service unavailable.

Question...

...but answer was different.

Mystery...

...but Scooby solved it.

Puzzle...

...but piece was missing.

Problem...

...but solution created more.

Issue...

...but subscription was cancelled.

Matter...

...but antimatter was cooler.

Concern...

...but worry was pointless.

Worry...

...but anxiety was worse.

Anxiety...

...but therapy was expensive.

Stress...

...but yoga class full.

Tension...

...but massage therapist booked.

Pressure...

...but tire was flat.

Strain...

...but muscle relaxer expired.

Burden...

...but back support broke.

Load...

...but washing machine broken.

Weight...

...but scale was lying.

Heaviness...

...but gravity was optional.

Lightness...

...but bulb burned out.

Brightness...

...but sunglasses were lost.

Darkness...

...but flashlight batteries dead.

Shadow...

...but sun went behind.

Shade...

...but tree was cut.

Twilight...

...but vampire was sparkly.

Dawn...

...but alarm didn't work.

Sunrise...

...but cloudy day predicted.

Sunset...

...but Netflix was buffering.

Night...

...but insomnia kicked in.

Day...

...but calendar was wrong.

Morning...

...but coffee maker broken.

Afternoon...

...but lunch was cancelled.

Evening...

...but dinner burned completely.

Midnight...

...but Cinderella was late.

Hour...

...but daylight savings confused.

Minute...

...but second hand stuck.

Second...

...but first was better.

Moment...

...but camera wasn't ready.

Instant...

...but coffee took forever.

Now...

...but then was different.

Then...

...but now is better.

Soon...

...but later was sooner.

Later...

...but now is good.

Eventually...

...but patience ran out.

Finally...

...but beginning was better.

End...

...but story kept going.

The ellipsis continues to create that perfect pause where the profound meets the absurd, making both more meaningful through their collision.Minimal Poetry Collection

Silence Speaks

...words fade...

The Last Drop

tears become...

Genesis

...light whispers...

Void

empty...full

Departure

...goodbye lingers

Dawn

darkness...surrender

Memory

...echoes remain...

Infinity

always...never...

Home

...heart returns

Truth

lies...then...

Ocean

waves...eternal...

Mirror

...self shatters

Time

...now vanishes

Stars

distance...longing

Breath

...life whispers...

Storm

chaos...then peace

Dreams

...morning steals

Fire

consume...become

Shadow

...follows always

Moon

...pulls tides

Rain

sky...weeps

Wind

...carries secrets

Stone

endures...

River

...flows toward

Mountain

stands...alone

Desert

thirst...teaches

Forest

...grows silent

Snow

covers...truth

Thunder

...speaks loudly

Lightning

instant...forever

Dawn

...hope returns

Dusk

day...surrenders

Night

...dreams begin

Morning

...new again

Sunset

...colors fade

Clouds

drift...disappear

Earth

...holds all

Sky

endless...blue

Horizon

...where meets

Path

...leads home

Journey

steps...toward

Destination

...never there

Walking

...one step

Running

...from what

Standing

still...moving

Sitting

...with silence

Lying

...truth hidden

Falling

...up toward

Rising

...from ashes

Breaking

...into pieces

Healing

...time mends

Hurting

...makes whole

Loving

...despite everything

Hating

...love disguised

Hoping

...against hope

Fearing

...the unknown

Knowing

...nothing certain

Learning

...to forget

Remembering

...what matters

Forgetting

...on purpose

Thinking

...too much

Feeling

...beyond words

Touching

...the untouchable

Seeing

...with heart

Hearing

...the unspoken

Tasting

...bitter sweet

Smelling

...memories return

Crying

...releases pain

Laughing

...despite tears

Smiling

...through sorrow

Frowning

...at joy

Sleeping

...while awake

Waking

...from dream

Dying

...to live

Living

...to die

Being

...and becoming

Becoming

...then being

Existing

...or not

Presence

...absence felt

Absence

...presence known

Here

...also there

There

...really here

Now

...then forever

Then

...now again

Forever

...just moment

Moment

...lasts forever

Always

...sometimes never

Never

...always maybe

Maybe

...definitely not

Definitely

...unsure though

Yes

...no perhaps

No

...yes hidden

Hello

...goodbye near

Goodbye

...hello waiting

Come

...go instead

Go

...stay longer

Stay

...leave soon

Leave

...return later

Return

...never left

Lost

...found within

Found

...lost again

Seeking

...what's here

Finding

...what's lost

Searching

...for home

Home

...everywhere now

Belonging

...to nowhere

Nowhere

...is somewhere

Somewhere

...is here

Everything

...becomes nothing

Nothing

...becomes all

All

...or none

None

...but this

This

...not that

That

...also this

Self

...meets other

Other

...reflects self

Together

...still alone

Alone

...never lonely

Lonely

...in crowds

Crowds

...one person

Person

...universe whole

Universe

...single breath

Breath

...connects all

Connection

...through separation

Separation

...creates unity

Unity

...holds difference

Difference

...same source

Source

...returns to

End

...new beginning

Beginning

...already ending

Circle

...no corners

Square

...round edges

Line

...curves eventually

Curve

...straightens out

Straight

...bends toward

Bend

...breaks free

Break

...mends stronger

Strong

...shows weakness

Weak

...finds strength

Strength

...in surrender

Surrender

...conquers all

Conquer

...lose yourself

Lose

...find everything

Find

...lose nothing

Nothing

...everything here

Here

...there too

Too

...just enough

Enough

...never plenty

Plenty

...still wanting

Want

...what have

Have

...what need

Need

...already here

Already

...not yet

Yet

...always now

Now

...never when

When

...doesn't matter

Matter

...doesn't when

Question

...answer itself

Answer

...questions more

More

...less than

Less

...more than

Than

...comparison ends

Ends

...begin again

Again

...first time

First

...last chance

Last

...first breath

Chance

...takes itself

Takes

...gives back

Gives

...receives more

Receives

...gives freely

Freely

...bound to

Bound

...set free

Free

...to choose

Choose

...no choice

Choice

...chooses you

You

...are me

Me

...are you

Are

...be becoming

Be

...or not

Not

...also yes

Also

...but not

But

...and still

And

...or maybe

Or

...and both

Both

...neither one

Neither

...both ways

Ways

...one path

One

...many faces

Many

...single heart

Single

...multiplied by

Multiplied

...divided into

Divided

...made whole

Whole

...broken open

Open

...closed circle

Closed

...opens wide

Wide

...narrow focus

Narrow

...expands beyond

Beyond

...within reach

Within

...outside bounds

Outside

...inside out

Inside

...outside in

In

...out beyond

Out

...in deeper

Deeper

...surface calm

Surface

...depths below

Below

...above ground

Above

...below sky

Ground

...sky meets

Meets

...parts ways

Parts

...whole story

Story

...silence tells

Tells

...without words

Without

...with everything

With

...without need

Need

...want nothing

Want

...need everything

Everything

...nothing more

Nothing

...everything less

Less

...more enough

More

...less plenty

Plenty

...enough never

Never

...always when

When

...never now

Now

...when always

Always

...never then

Then

...always now...

Silence Speaks

...words fade...

The Last Drop

tears become...

Genesis

...light whispers...

Void

empty...full

Departure

...goodbye lingers

Dawn

darkness...surrender

Memory

...echoes remain...

Infinity

always...never...

Home

...heart returns

Truth

lies...then...

Truth

lies...then...

Ocean

waves...eternal...

Mirror

...self shatters

Time

...now vanishes

Stars

distance...longing

Breath

...life whispers...

Storm

chaos...then peace

Dreams

...morning steals

Fire

consume...become

Shadow

...follows always

Moon

...pulls tides

Rain

sky...weeps

Wind

...carries secrets

Stone

endures...

River

...flows toward

Mountain

stands...alone

Desert

thirst...teaches

Forest

...grows silent

Snow

covers...truth

Thunder

...speaks loudly

Lightning

instant...forever

Dawn

...hope returns

Dusk

day...surrenders

Night

...dreams begin

Morning

...new again

Sunset

...colors fade

Clouds

drift...disappear

Earth

...holds all

Sky

endless...blue

Horizon

...where meets

Path

...leads home

Journey

steps...toward

Destination

...never there

Walking

...one step

Running

...from what

Standing

still...moving

Sitting

...with silence

Lying

...truth hidden

Falling

...up toward

Rising

...from ashes

Breaking

...into pieces

Healing

...time mends

Hurting

...makes whole

Loving

...despite everything

Hating

...love disguised

Hoping

...against hope

Fearing

...the unknown

Knowing

...nothing certain

Learning

...to forget

Remembering

...what matters

Forgetting

...on purpose

Thinking

...too much

Feeling

...beyond words

Touching

...the untouchable

Seeing

...with heart

Hearing

...the unspoken

Tasting

...bitter sweet

Smelling

...memories return

Crying

...releases pain

Laughing

...despite tears

Smiling

...through sorrow

Frowning

...at joy

Sleeping

...while awake

Waking

...from dream

Dying

...to live

Living

...to die

Being

...and becoming

Becoming

...then being

Existing

...or not

Presence

...absence felt

Absence

...presence known

Here

...also there

There

...really here

Now

...then forever

Then

...now again

Forever

...just moment

Moment

...lasts forever

Always

...sometimes never

Never

...always maybe

Maybe

...definitely not

Definitely

...unsure though

Yes

...no perhaps

No

...yes hidden

Hello

...goodbye near

Goodbye

...hello waiting

Come

...go instead

Go

...stay longer

Stay

...leave soon

Leave

...return later

Return

...never left

Lost

...found within

Found

...lost again

Seeking

...what's here

Finding

...what's lost

Searching

...for home

Home

...everywhere now

Belonging

...to nowhere

Nowhere

...is somewhere

Somewhere

...is here

Everything

...becomes nothing

Nothing

...becomes all

All

...or none

None

...but this

This

...not that

That

...also this

Self

...meets other

Other

...reflects self

Together

...still alone

Alone

...never lonely

Lonely

...in crowds

Crowds

...one person

Person

...universe whole

Universe

...single breath

Breath

...connects all

Connection

...through separation

Separation

...creates unity

Unity

...holds difference

Difference

...same source

Source

...returns to

End

...new beginning

Beginning

...already ending

Circle

...no corners

Square

...round edges

Line

...curves eventually

Curve

...straightens out

Straight

...bends toward

Bend

...breaks free

Break

...mends stronger

Strong

...shows weakness

Weak

...finds strength

Strength

...in surrender

Surrender

...conquers all

Conquer

...lose yourself

Lose

...find everything

Find

...lose nothing

Nothing

...everything here

Here

...there too

Too

...just enough

Enough

...never plenty

Plenty

...still wanting

Want

...what have

Have

...what need

Need

...already here

Already

...not yet

Yet

...always now

Now

...never when

When

...doesn't matter

Matter

...doesn't when

Question

...answer itself

Answer

...questions more

More

...less than

Less

...more than

Than

...comparison ends

Ends

...begin again

Again

...first time

First

...last chance

Last

...first breath

Chance

...takes itself

Takes

...gives back

Gives

...receives more

Receives

...gives freely

Freely

...bound to

Bound

...set free

Free

...to choose

Choose

...no choice

Choice

...chooses you

You

...are me

Me

...are you

Are

...be becoming

Be

...or not

Not

...also yes

Also

...but not

But

...and still

And

...or maybe

Or

...and both

Both

...neither one

Neither

...both ways

Ways

...one path

One

...many faces

Many

...single heart

Single

...multiplied by

Multiplied

...divided into

Divided

...made whole

Whole

...broken open

Open

...closed circle

Closed

...opens wide

Wide

...narrow focus

Narrow

...expands beyond

Beyond

...within reach

Within

...outside bounds

Outside

...inside out

Inside

...outside in

In

...out beyond

Out

...in deeper

Deeper

...surface calm

Surface

...depths below

Below

...above ground

Above

...below sky

Ground

...sky meets

Meets

...parts ways

Parts

...whole story

Story

...silence tells

Tells

...without words

Without

...with everything

With

...without need

Need

...want nothing

Want

...need everything

Everything

...nothing more

Nothing

...everything less

Less

...more enough

More

...less plenty

Plenty

...enough never

Never

...always when

When

...never now

Now

...when always

Always

...never then

Then

...always now...

3 or less words making profound poetry add ellipsis to end or beginnning and add a title

Edit
Minimal Poetry Collection
Document
Minimal Poetry Collection ## Silence Speaks ...words fade... ## The Last Drop tears become... ## Genesis ...light whispers... ## Void empty...full ## Departure ...goodbye lingers ## Dawn darkness...surrender ## Memory ...echoes remain...

Here's a collection of minimal poetry pieces, each using three words or fewer with ellipses to create profound emotional resonance. The brevity forces each word to carry maximum weight, while the ellipses suggest infinite possibility in the spaces between.

Retry
N

200 more

Edit
Minimal Poetry Collection
Document · Version 2
Truth lies...then... ## Ocean waves...eternal... ## Mirror ...self shatters ## Time ...now vanishes ## Stars distance...longing ## Breath ...life whispers... ## Storm chaos...then peace ## Dreams ...morning steals ## Fire consume...become

I've expanded the collection to over 200 minimal poems. Each one uses three words or fewer with ellipses to create profound moments of reflection. The poems explore themes of existence, time, consciousness, relationships, and the human condition through the power of brevity - where every word carries immense weight and the ellipses suggest infinite possibility in the spaces between thoughts.

Retry
N

add 200 more but add humor

Edit
Minimal Poetry Collection
Document · Version 3
Always ...never then ## Then ...always now ## Now ...when always ## When ...never now ## Monday ...already Friday? ## Friday ...feels like Monday ## Coffee ...or death ## Death ...without coffee ## WiFi ...life's meaning ## Meaning ...WiF

I've added 200 more minimal poems with humor! These range from relatable modern struggles (Monday blues, Netflix addiction, missing socks) to absurd progressions that spiral into comedic territory. The humor comes from the unexpected connections, the mundane made profound, and the way simple observations about daily life become existential sock philosophy. The ellipses create perfect comedic timing, letting each punchline land with maximum impact.